Solveig:
Best!
Alan P. Garfinkel
2007

ARCHAEOLOGY and ROCK ART
of the
Eastern Sierra and
Great Basin Frontier

ALAN P. GARFINKEL

with a foreword by

Michael J. Moratto

Maturango Press • Ridgecrest, California

MATURANGO MUSEUM

First printing: Maturango Press, Summer 2007
Maturango Museum Publication No. 22

Edited by members of the Publications Committee,
Elizabeth Babcock, chairman

Cover art, "Portal of Power," by Cathy Crandall, *tofuden2001@yahoo.com*

ISBN 0-943041-16-3

MATURANGO
PRESS

For my wife, Leanne, my children, Jason, Max, and Hannah,
and my friends Richard Blalock, Dr. Donna Kono,
and Julia Schley.

Thank you for keeping the dream alive and
for your support throughout the completion of that vision.

Table of Contents

List of Figures

List of Tables

Foreword

Archaeology and Rock Art… advances our knowledge of early human societies and their ways of life in a remote and little-known part of eastern California. This book is adapted from Alan Garfinkel's Ph.D. dissertation (University of California, Davis, 2005), based on extensive library research and field investigations beginning nearly 30 years ago. As the dissertation title indicates, the study is concerned with *Linguistic Archaeology: Prehistoric Population Movements and Cultural Identity in the Southwest Great Basin and Far Southern Sierra Nevada.* Although the subject matter may at first appear to be recondite or arcane, in reality it is fascinating. Dr. Garfinkel is to be commended for his lucid presentation of research goals, methods, and results as well as for showing how his findings relate to larger issues of ecological and anthropological interest.

Geographically the present work encompasses a 35-mile (56-kilometer) segment of the Pacific Crest Trail (PCT) traversing parts of the Kern Plateau and Scodie Mountains. Located approximately 25 miles (40 kilometers) west of U.S. Highway 395, this is an area of remarkable natural diversity. The rolling terrain of the Kern Plateau, with its granitic ridges and domes rising 5000-8000 feet (1525-2440 meters) above sea level, is the southern tip of the Sierra Nevada. The Scodie Mountains, however, are northern outliers of the Tehachapi Range—part of the geologically complex Transverse Ranges. Within the study area, the highland crest followed by the PCT sharply divides the semi-arid to subhumid Kern Plateau from the xeric (dry) Mojave Desert and Great Basin. Characteristic vegetation includes Montane Meadow, Mixed Conifer, Single-leaf Piñon/Utah Juniper, and Joshua Tree series. Among the local resources of economic value to native peoples are extensive stands of piñon pines and abundant game animals, notably bighorn sheep, deer, pronghorn, and a host of smaller creatures.

The spatial focus of the book thus coincides with the nexus of four major geomorphic provinces: the Sierra Nevada, Transverse Ranges, Mojave Desert, and Great Basin. In addition, the plateau is drained by the north and south forks of the Kern and linked by that river to the southern San Joaquin Valley. It is not surprising, therefore, that this juncture of the California and Great Basin culture areas is an ideal setting for archaeological research to test a wide range of hypotheses dealing with local and regional prehistory and human ecology.

Paralleling the area's natural diversity is a mosaic of native languages and ethnic territories, not unlike the myriad speech communities found in other parts of aboriginal California. The Kern Plateau was inhabited mainly by speakers of Tubatulabal, a distinct member of the Uto-Aztecan language family. To the north were the Owens Valley Paiute; to the east, in the Coso Range and beyond, lived the Panamint Shoshone; and to the south was Kawaiisu country. Bordering the Tubatulabal on the west were various Yokuts tribelets in the Sierra foothills and San Joaquin Valley. Significantly, Paiute, Shoshone, and Kawaiisu all belong to the Numic subfamily of Uto-Aztecan and are thus distantly related to Tubatulabal. The Yokutsan family, however, is a division of the California Penutian stock and has no known linguistic ties to Tubatulabal.

While speakers of Penutian languages once occupied much of central California, Numic peoples claimed the entire Great Basin except for a relatively small area near Lake Tahoe and on the adjacent eastern Sierra front. What makes such linguistic detail pertinent here is that questions about the origins, age, dispersion, and adaptive practices of Numic peoples have for many decades been at the heart of anthropological debate and research in the Great Basin.

Garfinkel has taken full advantage of the rich archaeological potentials afforded by the eastern Sierra and Great Basin frontier. His research is designed to shed light on several themes and problems of enduring interest to scholars working in the region. Foremost, he delves into linguistic prehistory and evaluates several alternative models that seek to account for the timing and dynamics of Numic territorial expansion. This examination, in turn, requires identifying past ethnic or linguistic groups by their archaeological "signatures"—always a daunting task with many assumptions inherent.

Garfinkel also examines, among other things, relationships among climatic change, environmental response, and cultural adaptations; impacts of aboriginal hunting on target wildlife species, especially bighorn sheep; the role of prestige in cultural evolution; and the use of mitochondrial DNA (mtDNA) to identify ethnic groups (actually, populations) in prehistory. Especially notable is his perceptive analysis of rock art and full integration of the results with other kinds of archaeological data to help meet the study's research objectives.

In addition to rock art, the material basis for this study includes data from 69 sites distributed among seven locations on the Kern Plateau and in the Scodie Mountains. Much of the information comes from sites along the PCT excavated by Garfinkel and his colleagues during the 1980s. More recently, state and federal agencies (especially the California Department of Transportation and the Naval Air Weapons Station, China Lake) have funded numerous archaeological projects nearby in the western Great Basin and northern Mojave Desert.

The research, carried out by such scholars as M. Basgall, R. Bettinger, M. Delacorte, R. Elston, A. Gilreath, M. Hall, W. Hildebrandt, K. McGuire, L. Reynolds, M. Sutton, D. Whitley, and W. Woolfenden, has added greatly to our knowledge of regional prehistory and cultural ecology. It is within this regional context that Garfinkel interprets the archaeological assemblages, rock art and evidence of diachronic land-use and economic practices within the PCT study area.

Garfinkel's analyses lead to a series of conclusions. First, the continuity of subsistence practices, toolstone use, and rock art styles during the past 2500 years indicates that the Tubatulabal cultural tradition—and, presumably, language—are of long standing in the South Fork Valley of the Kern Plateau interior.

Second, from about 500 B.C. until A.D. 600, non-Tubatulabalic, "pre-Numic" peoples seem to have inhabited lowland areas of the adjacent desert (east of the Kern Plateau), where they hunted artiodactyls, gathered brown-cone piñon nuts, and exploited riparian resources.

Third, in the desert, Numic populations began to replace established, pre-Numic people by ca. A.D. 600. This change is shown archaeologically by the advent of the bow and arrow and new economic modes evinced by upland piñon camps, lowland seed-gathering stations, and other special-purpose sites. Next, the interval from ca. A.D. 600-1300 (Haiwee Period) witnessed dramatic environmental shifts, particularly the severe droughts of the "Medieval Climatic Anomaly" (MCA) after ca. A.D. 890 and concomitant cultural adjustments. It was during this period that many established (pre-Numic) settlements were abandoned and production of Coso Representational Style rock art ceased. After A.D. 1300 (Marana Period), any remaining pre-Numic people evidently were absorbed or replaced as a consequence of the Numic expansion.

Garfinkel asks how Numic groups were able to supplant populations that had lived successfully in the western Great Basin for hundreds or thousands of years. He takes up this question by evaluating three possible scenarios of Numic prehistory. His data seem most compatible with the "economic displacement" model proposed in 1982 by Bettinger and Baumhoff. This model posits that pre-Numic societies were characterized by low population densities of "travelers" who invested a lot of energy in hunting large game and acquiring high-ranked plant resources (e.g., ripe piñon nuts). By contrast, Numic "processors" had higher population densities and spent more time gathering low-ranked foods such as hard seeds and small game. Thus, Numic social organization and economic modes might have conveyed a significant competitive advantage over pre-Numic peoples during times of environmental change and ecological stress, such as the decline or extermination of big game caused by over-hunting, drought, or disease.

To sum up, this book is a major contribution to California/Great Basin archaeology and several of its subfields. Garfinkel has synthesized a great deal of previously unpublished and covert archaeological information. He has argued cogently for long-term continuity *in situ* of the Tubatulabal and for replacement of pre-Numic by Numic populations east of the Sierra, thereby elucidating the larger domain of linguistic prehistory.

The book also serves as a case study in cultural ecology that examines the success or failure of different adaptive strategies in the context of environmental change over time. Finally, and perhaps most significantly, the author's analysis and interpretation of rock art are truly exemplary. Although the vast California/Great Basin literature includes myriad reports of "archaeological" *or* rock-art research, it is rare to find the two so thoroughly—and so effectively—integrated. In this sense and many others, Garfinkel has produced a book worthy of emulation.

Michael J. Moratto
Senior Archaeologist
Applied EarthWorks, Inc.

Acknowledgements

This book may be a bit unusual in that its preparation was interrupted by a two-decade hiatus. My original effort produced a draft in 1983, and I completed the present research in 2005. A few challenges interrupted the initial effort, not the least of which was the untimely death in 1983 of Dr. Martin Baumhoff, the chair of my original dissertation committee. With Dr. Baumhoff's passing and a series of other complications, I left active studies at the university only to reenter the field of archaeology nearly 20 years later.

I was very fortunate to have helpful friends who would support my renewed initiative and provide me with a needed refreshing on the current status of the profession. These include some of my lifelong colleagues: Adelle Baldwin, Jeanne Day Binning, William Hildebrandt, Rob Jackson, Russell Kaldenberg, Valerie Levulett, Kelly McGuire, Tom Origer, John Romani, Mark Sutton, Robert Schiffman, David Whitley, and Robert Yohe.

Several other people furnished helpful comments, enlightenment, and support along the way including Mark Allen, Don Austin, Mark Basgall, Tim Carpenter, Carol Carroll, Michael Delacorte, Jelmer Eerkens, Mark Faull, Jill Gardner, Sarah Gassner, Amy Gilreath, Russell Kaldenberg, Don Laylander, Geron Marcom, Wendy Nelson, Barry Price, Steve Ptomey, Bill Ray, Alexander Rogers, Fran Rogers, Jeff Rosenthal, Jane Sellers, Craig Skinner, Nathan Stephens, Claude Warren, John Whitehouse, Eric Wohlgemuth, Elva Younkin, and David Zeanah. All of these individuals were most patient in listening to my thoughts and offering insights at various points during the present effort; for their help I am most thankful.

A special thank you goes out to my newfound friends, J. Kenneth and Anna Lu Pringle, who always offered their positive perspective and support when I was in need of re-energizing and reassurance.

A great measure of thanks is also offered to those people who worked in the field and laboratory collecting and analyzing the data from the Pacific Crest Trail archaeological projects. I am grateful to all who assisted in preparing the final reports. From the initial Scodie Mountains segment those persons included Richard Ambro, Michela Crist, A. Van Eggers, Ronald May, Clement Meighan, Tom Origer, Billy Peck, Dudley Varner, Sonja Wilson, and Patti Wermuth. Follow-up studies at Ker 1286 in the Scodie Mountains segment were completed by Mark Basgall, James Bennyhoff, Paul Bouey, Michael Delacorte, Suzanne Griset, Rob Jackson, Kelly McGuire, Tod Ruhstaller, and G. James West.

For the Morris Peak and Lamont Meadows segments, those working to complete the study included Cynthia Adams, Stephen Andrews, Valerie Asakawa, Gayle Bacon, Stephen Bass, Leann Brooks, Michael Buchak, Franz Cilensek, Charles Collings, Lisa Dillinger, Richard Hughes, Kelly McGuire, Ronald May, Sue Ann Monteleone, Tom Origer, Martha Proctor, David Rhode, Lorna Rocha, Robert Schiffman, Jeannette Schulz, and Matthew West.

Individuals who were responsible for completing the Bear Mountain study included Gayle Bacon, Mark Basgall, Stephen Bass, Michael Bowers, Tony Drake, Bob Pfiel, Richard Hughes, Robert Jobson, Paul Keyser, Valerie Levulett, Helen McCarthy, Kim McGuire, Ronald May, Christa Maxfield, Sheila Mone, Sue Ann Monteleone, Tom Origer, Zora Parkevitch-Tammer, David Rhode, Tod Ruhstaller, Laurie Swenson, Wendy Waldron, Patti Wermuth, and G. James West.

For the Kennedy Meadows and Rockhouse Basin segments, the following people were involved: Gayle Bacon, Mark Basgall, James Bennyhoff, Paul Bouey, Kevin Clark, Suzanne Griset, Bill Hildebrandt, Liz Honeysett, Richard Hughes, Robert Jackson, Paul Keyser, Kelly McGuire, Nancy Ridgway, Todd Ruhstaller, Dolly Stengl, Laurie Swenson, and Patti Wermuth.

The many individuals who helped with the Rockhouse Basin Segment data recovery study included Rebecca Anastasio, Frank Asaro, James Bard, Mella Bellman, David Bibler, Paul Bouey, James Buckley, Colin Busby, Douglas Davy, Mary Davy, Raymond Dezanni, Michael Foley, Amy Gowan, Robert Harmon, Axel Hoffman, Rob Jackson, Melissa Kennard, Larry S. Kobori, Annemarie Meike, Helen Michael, Patricia Ogrey, Howard Schorn, Michael Siskin, Melody Tannam, Richard Weaver, Patti Wermuth, Thomas Whitman, Beverly Wigley, and Charley Wigley.

Special thanks are additionally offered to Mark Basgall, Michael Delacorte, Jelmer Eerkens, Mark Faull, Amy Gilreath, Bill Hildebrandt, Geron Marcom, Mark Sutton, Audrey Williams, Robert Yohe, and David Zeanah for sharing details of their unpublished research. Helen Clough and Eric Montizambert represented the Bureau of Land Management and facilitated the contractual elements for several of the Pacific Crest Trail projects. Glen Caruso, Steve Fosberg, Paul Short, and Sonia Tamez, of the U.S. Forest Service, helped manage the agreements for cultural studies on other Pacific Crest Trail segments.

I express special gratitude to Lyn Meckstroth of Applied Earthworks, Inc., for the illustrations accompanying this book. I am also indebted to the members of my dissertation committee: Drs. Conrad Bahre, Michael Barbour, and Michael Moratto. Drs. Bahre and Barbour co-chaired my committee and represented the University of California Davis. Dr. Bahre's background in cultural geography and Dr. Barbour's expertise in the environmental ecology of the Sierra Nevada were invaluable. Dr. Moratto served as an outside committee member offering his expertise on California and Great Basin linguistic prehistory and archaeology and providing good, commonsense thinking — dramatically improving the scientific merits and precision of this research. Additionally, during several years of Friday night get-togethers, my friends Richard Blalock, Dr. Donna Kono, and Julia Schley never let me forget my dream of finishing my dissertation and obtaining my doctorate. Their good-natured nudging resulted in my urge to return to the discipline and complete my unfinished task.

I would be remiss if I had forgotten to recognize my family members and relatives who prayed with me and humored me through my efforts. A giant debt of gratitude goes out to my mother, Louise Garfinkel-Santangelo, stepdad, John Santangelo, and my wife, Leanne Gold. Special recognition is necessary for my in-laws, Martha, John, and Bradley Maddox, and my aunt, Judith Wren, who aided me in great measure both financially and emotionally throughout our recent move to Bakersfield. Our children, Jason, Max, and Hannah, also somehow survived the toils of this extended effort, and to them I am thankful for putting up with such an enormous task.

A special thanks must also go to all my colleagues at the Maturango Museum, most especially to my diligent and patient editor Liz Babcock and other members of the editorial team, Jean Bennett and Ruth Dietrich. To all those I have remembered here and to any I have inadvertently overlooked, I express my heartfelt thanks in enabling me to complete this book.

A.P.G.
Bakersfield, California
November 15, 2006

Chapter 1
Introduction

Scope and Purpose

The Eastern Sierra and Great Basin frontier is an area of startling contrasts. Only 50 miles wide and 100 miles long, it is located at the southernmost and lowest portion of the great Sierra Nevada mountain chain. It is a fascinating land of raw and powerful beauty that incorporates the headwaters of the south fork of the Kern River and the rugged Kern Plateau. In a drop of more than 5000 feet, you can travel from the well-watered uplands of piñon-juniper woodland along the crest of the Sierra down through a dramatic forest of Joshua trees. As you wind down the eastern scarp, you catch a brief glimpse of Great Basin sage only to arrive on the arid valley floor. There you are greeted by a panorama of desert mountains, dry lakes, and volcanic lava flows. Thin, pungent, grey-green shrubs pepper the landscape. These are the kings of the Mojave Desert plants — resilient creosote bushes.

This is a mysterious land that takes time to truly know. Vistas are painted in sharp, strong colors, and the geologic forces, that shaped the landscape, are here revealed for all to know. The area has many layers, and each must be digested slowly and savored to appreciate their intimate connections. Those relationships between plants, animals, rocks, and people, are part of the story told within the pages of this book.

Yet the focus of this narrative is also a blend, a mix between dirt archaeology, the study of religious iconography, and linguistics, with a small dash of population genetics thrown in for good measure. Some scientists have labeled that subject linguistic archaeology, and that is about as good a title as one can get. Language reconstructions trace the history and prehistory of Indian languages and peoples identifying their possible patterns of movement, spread, and replacements over time. A select group of prehistorians have chosen this subject as their specialty. These scholars take the models of their scientific partners and test them.

The interface of the Eastern Sierra and the Great Basin provides an excellent outdoor laboratory for this type of research. Anthropologists have been impressed with the diversity of native languages spoken in eastern California. No less than four distinct Indian groups, all having different languages, are known historically. These peoples formed a mosaic with various territories mapped over this rather small area. In the Isabella Basin and Kern River uplands, the Tubatulabal (piñon pinenut eaters) lived. Their tongue was related to a large language category known as Uto-Aztecan. Other speech communities connected with this branch were spoken throughout the Great Basin, the American Southwest, and into northern Mexico. To the south and east were speakers of several Great Basin Shoshonean or Numic languages. In the Tehachapi Mountains were the Kawaiisu, around Little Lake and the Coso Range were the Panamint Shoshone, and in the Owens Valley were the Northern Paiute.

In the rest of the Great Basin fanning out from eastern California were other related languages. These languages extend in a vast triangle with its pointed end in the southern Sierra Nevada and its base along the Rocky Mountain chain. This great area includes the interior Great Basin, the Snake River Plain, and part of the Colorado Plateau. Linguists conclude from this distribution and closer study of the form and character of these languages that in recent times (less than a thousand years ago) a major migration occurred out of the study area that ultimately covered the entire Desert West. This migration is known as the "Numic expansion." A majority of linguists have come to believe that the area of origin for a language group is correlated with the region inhabited by its greatest diversity. If such were the case, somewhere within the vicinity of our study area was the hearth where the Numic languages originally developed. Therefore, one of the principal research questions that tugs at prehistorians working in this region is the way to best interpret dynamic cultural changes in contrast to relatively stable patterns of cultural longevity. In some portions of the study area, archaeologists can trace culture history back for thousands of years with little apparent change in the native lifeways. Yet in other nearby areas there appear to be radical shifts in the lifeways of the people over that same time span.

This book asks the question and tries to ferret out whether these changes are associated with the movements of particular ethnic/language groups. We go one step further in considering what distinctive archaeological patterns might conform to ethnic/linguistic change, in-migration, and population replacement.

Seven archaeological studies, along the crest of the Eastern Sierra form the basis for much of the evidence reviewed in this book. These studies were conducted in the 1980s in order to lessen the impacts to fragile prehistoric sites that could have been destroyed by the construction of the Pacific Crest Trail. The evidence includes analysis of 69 archaeological sites, excavation results from 54 sites, 475 obsidian hydration dating measurements, and 28 radiocarbon dates. Additional data (the dating and styles of rock paintings and rock drawings, mitochondrial DNA analyses, burial patterns, regional and site-specific obsidian hydration dating patterns, toolstone material use, dietary patterns, and distribution of time-diagnostic artifacts) were gathered from archaeological studies in the southwestern corner of the Great Basin (Indian Wells Valley, Rose Valley, and the southern Owens Valley in eastern Kern and southern Inyo counties). All of these data are considered with regard to whether they support models favoring population replacement or cultural continuity.

This study integrates historical, linguistic, and archaeological studies to model the movements of speakers of Numic and Tubatulabal languages along the margin of the southwestern Great Basin. Data from the seven archaeological studies in the far southern Sierra Nevada and Scodie Mountains are reviewed (Figures 1.1, 1.2, and 1.3). Two types of models are considered: (1) those favoring *in situ* development of the Numic pattern, and (2) those hypothesizing movement, expansion, and population replacement.

This chapter briefly (1) introduces the problem of recognizing archaeologically the territorial boundaries of ethnic/linguistic groups, (2) traces the history of Great Basin linguistic models and the debate concerning Numic expansion, and (3) evaluates the implications of models of population shifts versus in-place culture change and continuity.

Anthropological Background

Since the beginnings of anthropology more than a century ago, prehistoric population movements have been a focus of study. Interest has of course waxed and waned, but researchers have often returned to the topic because of its central place in anthropological studies. Closely tied to the topic of population movements are the relationships of material culture to other aspects of human behavior and the link between language and ethnicity.

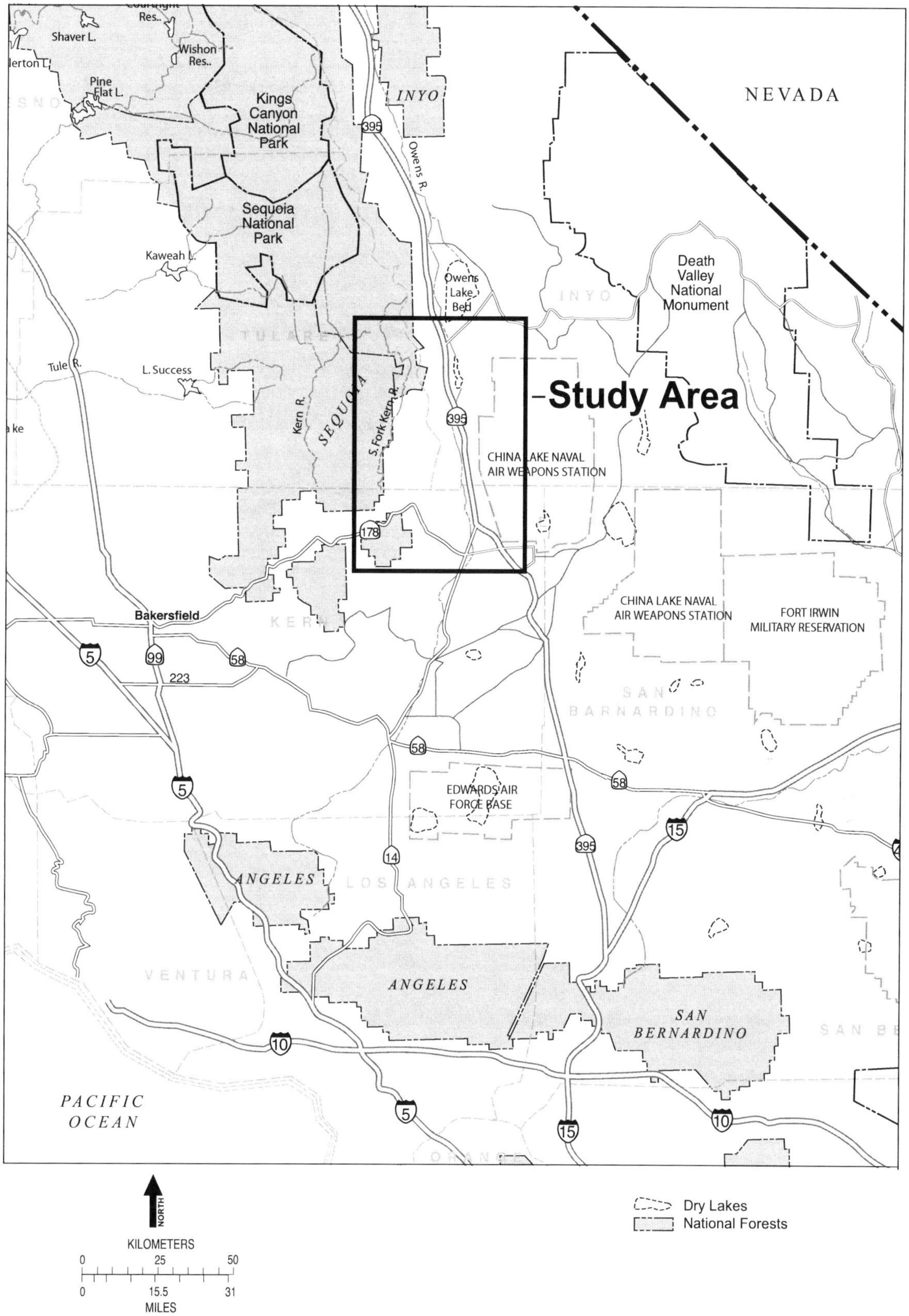

Figure 1.1 General Location of Study Area.

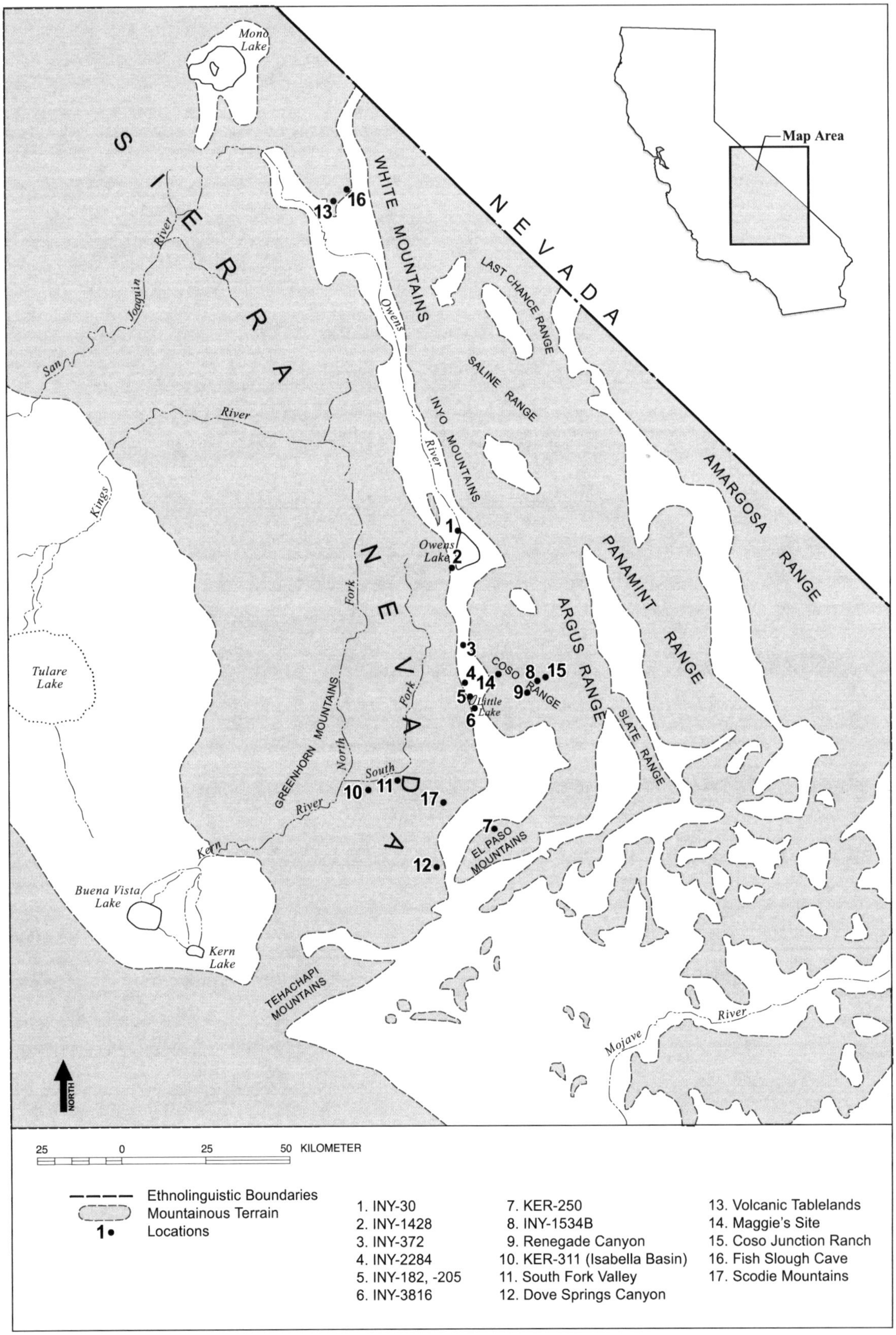

Figure 1.2 Important Locations in the Region.

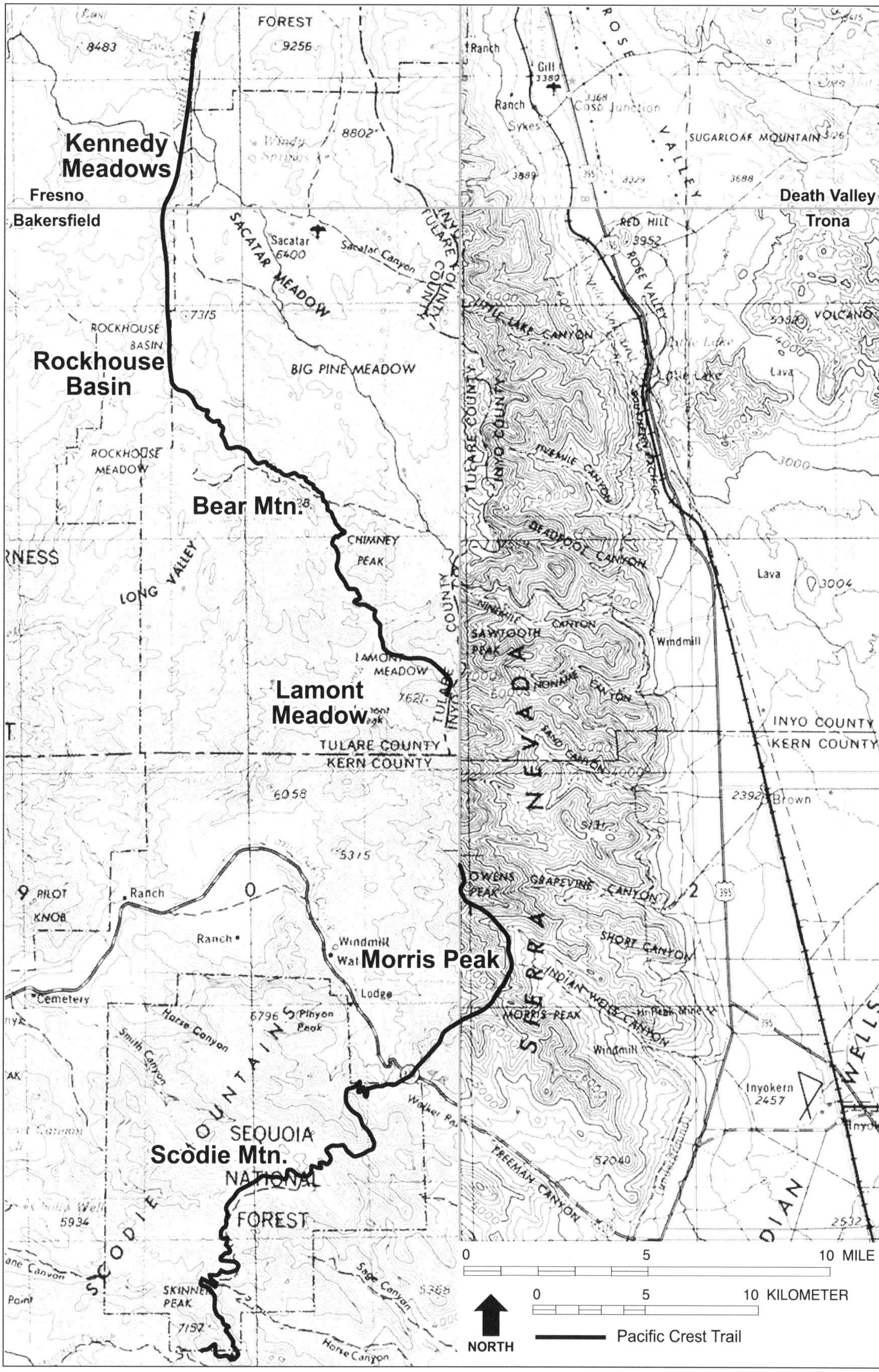

USGS 1:250 000 Bakersfield, Death Valley, Fresno, and Trona, CA

Figure 1.3 Map of Pacific Crest Trail Segments and Study Area.

With every theoretical and methodological innovation have been concomitant refinements in cultural studies of sociopolitical organization, subsistence-settlement, and cosmology. All eventually depend on our ability to study the movement and spread of prehistoric populations, ethnic shifts, and language change. Yet over the last century, debate has continued to focus on when and how native peoples came to have their historical distributions. In the 1980s and 1990s the topic again came to the forefront, receiving attention worldwide (DeAtley and Findlow 1984; Ehret 1988; Foster 1996; Hughes 1992; Lindstrom 1996; Nash 1996; Renfrew 1987; Renfrew et al. 2000; Roberts et al. 1995).

Nature of the Problem

Tracing the movements of languages and people through time and space is a difficult undertaking, since no necessary relationship exists between language, ethnicity, and material culture (cf. Foster 1996; Hughes 1992; Kroeber 1955:104). People who speak closely related languages can exhibit dramatic differences in material culture, and groups can have similar material cultures yet speak vastly different languages. But, given these well-known facts, anthropologists can still develop theoretically sophisticated proposals based on the available evidence that can be challenged, refined, and debated. Although far from definitive, linguistic archaeology can illuminate the past by serving as a source for hypothesis generation and making us aware of the possible prehistory of a given area.

Attempts to infer ethnolinguistic attributes and meaning from archaeological patterns are by their very nature problematic. Relating prehistoric assemblages to particular linguistic units is more often informed guesswork than systematic scientific inquiry and is subject to a great range of interpretive challenges. Yet, when we speak of large-scale language movements *among hunter-gatherers*, we are *normally* talking about language shifts that occur because their speakers change their geographical range (cf., Bellwood 1985, 1997; Bettinger and Baumhoff 1982; Ehret 1976; Renfrew 1987). This is language replacement writ large, based on population movements where the speakers of the indigenous language often do not retain their former territory. Language spread often involves movements of people into already populated areas; therefore, some competitive advantage must ordinarily be assumed.

Ethnic Groups and Their Archaeological Correlates

Ethnic groups are usually defined by the fact that they share a common language and are biologically self-perpetuating (Barth 1969). Notwithstanding the long-running debate on exactly which attributes should best be interpreted as having stylistic rather than purely technofunctional significance (Binford 1968, 1986; Sackett 1985, 1986), style may be defined as "the formal variation in material culture that transmits information about personal and social identity" (Weissner 1983:256). As such, artifact styles have the potential to aid in the study of ethnic boundaries, intergroup competition, and population displacements

Stylistic attributes or cultural elements argued as *not purely functional* are useful in distinguishing ethnic groups archaeologically (Adovasio 1986; Wiessner 1983, 1985). Material culture elements serving in a religious or ideological context can serve especially well as proxy indicators of ethnicity (e.g., rock art, burial patterns or iconographic pottery designs). When stylistic elements simultaneously occur in space and time, the argument for ethnic distinction becomes compelling (Hodder 1982; Simpson 1988). This distinction may correspond to any of a number of sociocultural levels (including family, clan, lineage, village, tribelet, territorial band, or ethnolinguistic group).

Hunter-Gatherer Territorial Boundaries

Much discussion has focused on whether hunter-gatherers hold and defend their core territories, and whether they exhibit exclusive ownership of resources within such territories (Kroeber 1925; Steward 1938, 1970). In practice, boundary maintenance in hunter-gatherer communities varies widely. On a worldwide basis we can recognize examples where simple foraging peoples do mark their territories and have distinctive cultural elements that serve as indicators of cultural interaction, ethnic-linguistic signatures, and tribal boundaries (R. Layton 1986; Peterson 1978:24–25; Weissner 1983).

Foragers' core territories may display high densities of particular stylistic elements and there may be sharp, well-defined dropoff shoulders at their boundaries (Hodder 1982). Mapping of such stylistic "group-marker elements" would establish their relative densities across the landscape. Simple distance-decay models would not be expected when ethnic distinctions and territoriality are involved. The scale of such distributions spatially will provide clues to levels of sociocultural distinction.

Historical Linguistic Models in the Great Basin and Study Area

Scholars continue to debate the prehistoric developments leading to the historical distributions of Uto-Aztecan (and specifically Numic) languages. Historical linguist Sydney Lamb (1958) first suggested that Northern Uto-Aztecans moved into the western United States from a homeland in northern Mexico about 5,000 years ago and subsequently diverged into four branches: Hopic, Numic, Takic, and Tubatulabal. Hopic was thought to have moved to the east while Tubatulabal remained in place.

Numic and Tubatulabal were believed to have separated in the western Great Basin and far southern Sierra Nevada, with Tubatulabalic enjoying some 2,000–3,000 years of in-place development. The division of Numic into its separate branches and individual languages is thought to have occurred within the last 1,000–2,000 years.

In terms of linguistic classification, Numic consists of three subdivisions: Western, Central, and Southern (Kroeber 1907, 1925). Each Numic subdivision comprises essentially two languages, one occupying a small area in eastern California and the other a large portion of the Great Basin (Figure 1.4). Applying the center-of-gravity principle (Sapir 1916), we can reasonably conclude that the Numic homeland lay in the area of greatest linguistic concentration. The Numic languages probably spread outward from this linguistic hearth in the southwestern Great Basin in the not-too-distant past.

Many scholars have concluded that Numic populations expanded into the Great Basin from this linguistic center $\leq$ 1,000 years ago (Foster 1996; Miller 1986; Moratto 1984: Chapter 11).

History of Debate Concerning Numic Expansion

During the 1960s and 1970s the archaeological community largely agreed on the above scenario, at least in acceptance of the expansion of Numic people. But issues of timing, as well as the mechanisms and geographical direction of the spread, were still controversial.

In the 1980s the tenuous consensus began to disintegrate, and in the mid-1990s a historical benchmark was set with the publication of the Madsen and Rhode monograph (1994). That work served to focus wide-ranging and lively discussions questioning the assumptions for Numic expansion and the nature of language shifts over time.

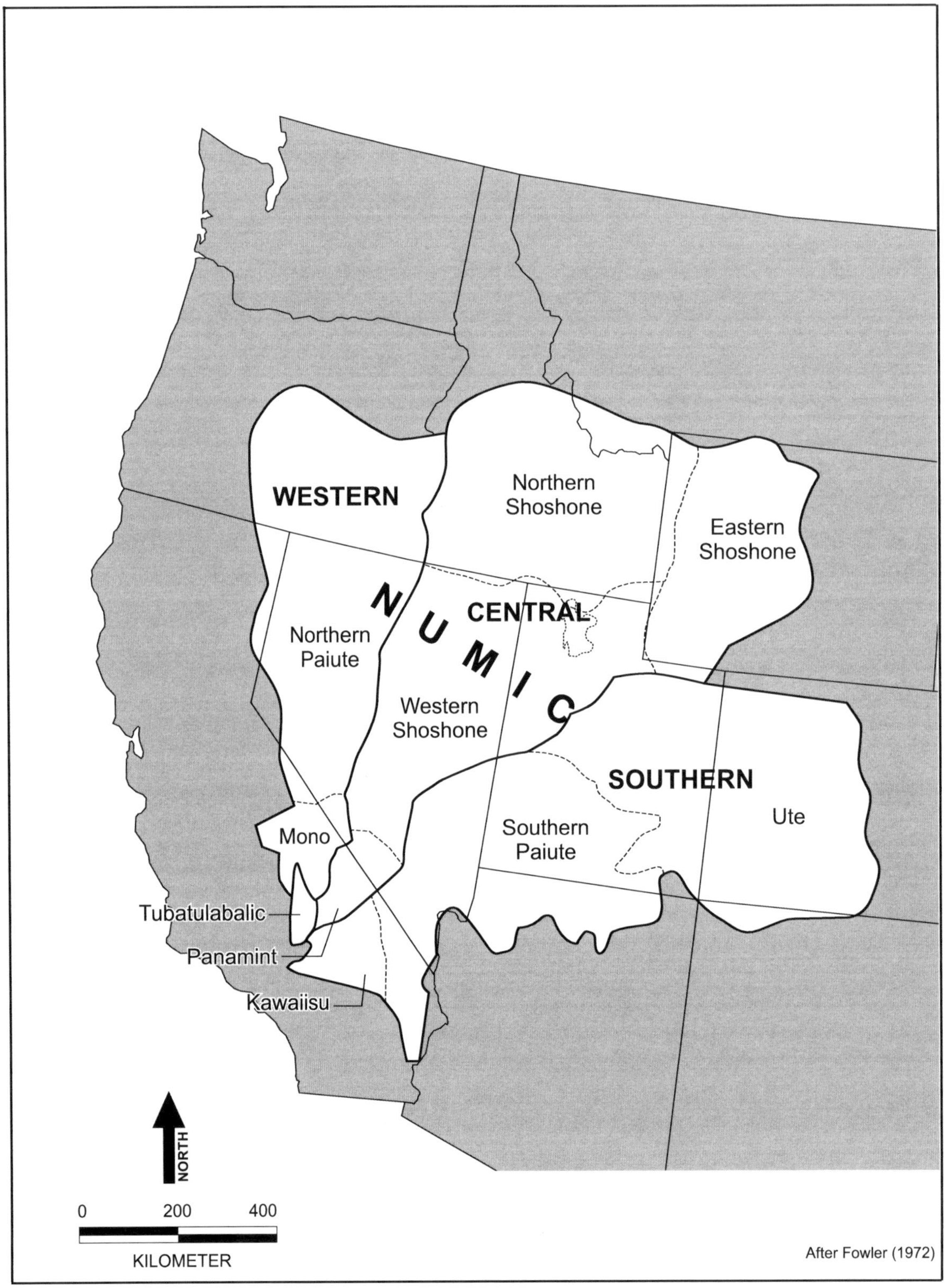

Figure 1.4 Distribution of Numic Languages, ca. A.D. 1850.

As discussed in more detail below, the diverse views on Numic prehistory fall into two groups: those accepting the replacement of pre-Numic populations by Numic peoples (Adovasio 1986; Bettinger 1994; Bettinger and Baumhoff 1982; Madsen 1994; Sutton 1986; 1987; Young and Bettinger 1992) and those favoring a lengthy *in situ* development of Numic groups within the Great Basin, including the study region (Aikens 1994; Aikens and Witherspoon 1986; Goss 1964, 1977; Grant et al. 1968; Pearson 2002; Whitley 1998).

Current Research

Even after more than a century of debate and a multitude of publications, scholars can agree only on the historical linguistic distribution of Numic and other Uto-Aztecan peoples. Disagreements pervade the discussions because competing models have yet to be rigorously and critically tested (Fowler 1972; Madsen 1975; Sutton 1986). The legitimacy of the tests has been questioned because the results have been equivocal. Also, anthropologists have not developed generally accepted means for tracing prehistoric population movements. There is little agreement on the way to identify the archaeological markers of language shifts. Neither is there clear midrange theory relating archaeological data to models of prehistoric population movements, language shifts, and ethnic spreads.

Additionally, cultural chronologies in the West are not definitive; hence they leave considerable room for argument about dating and the associations of cultural sequences with particular cultural identities. Chronologies based upon projectile points are fairly well established, yet the dating of key series and the exact definition of certain types are still subject to considerable disagreement (e.g., the Humboldt and Pinto problems).

Great Basin archaeology is often represented by surface materials. Dating such sites has at times been difficult. The means of doing so are still subject to issues of resolution and accuracy (e.g., the use of obsidian hydration and time-sensitive artifacts for dating). Disagreements still relate to basic questions of diet, settlement pattern, site function, and age of archaeological patterns, including such longtime standards as "the nature and antiquity of piñon exploitation for eastern California" (Bettinger 1976; Bettinger and Baumhoff 1982; Hildebrandt and Ruby 2000; McGuire and Garfinkel 1976; Reynolds 1996).

Alternative Models of Numic Prehistory

Alternative archaeological reconstructions and varying interpretations of the linguistic data have led to various reconstructions of population movements in the Great Basin. The Numic expansion and the proposal that Numic languages are of late prehistoric age in the Great Basin have generated a great deal of discussion and debate.

The timing and direction of Numic movements have also been the subject of considerable debate (Aikens 1994; Aikens and Witherspoon 1986; Bettinger and Baumhoff 1982; Goss 1977; Sutton 1987, 1991). It seems that around the peripheries of the Great Basin, archaeologists and linguists have less of a problem distinguishing and tracing the movements, timing, and archaeological features of the Numic intruders and their pre-Numic progenitors (see Grayson 1994; T. Layton 1985). The problem lies with the identification of unique elements of the different patterns of the Numic pioneer stream.

Although many scholars have presented evidence to support the replacement hypothesis — that is the replacement of pre-Numic populations by Numic (Bettinger and Baumhoff 1982; Sutton 1987; Young and Bettinger 1992) — others have argued for lengthy occupation by Numic peoples in various locations within the Great Basin, including the general study area (Aikens 1994; Aikens and Witherspoon 1986; Goss 1977; Grant et al. 1968; Pearson 2002; Whitley 1998).

Description of Study Area and Archaeological Data

Seven extensive archaeological studies have been conducted along and near the crest of the far southern Sierra Nevada on the Kern Plateau and in the Scodie Mountains (Ambro et al. 1981; Bard et al. 1985; Garfinkel et al. 1980, 1984; McGuire 1981, 1983; McGuire and Garfinkel 1980). These investigations were prompted by the need to mitigate impacts from the construction of the Pacific Crest Trail on a series of archaeological sites (Figures 1.5–1.10). Data were obtained from the study of 69 individual archaeological sites; excavation results from 54 cultural deposits; 475 obsidian hydration measurements; and 24 radiocarbon dates. It is rather curious that since the mid 1980's little additional research has been conducted in this area apart from limited surface surveys and site recording (Scott and Obrien 1991; W & S Consultants 1999).

Synthesis, reevaluation, and a detailed examination of these and other related data form the basis for the present study. Also relevant is the consideration of other archaeological research in the southwestern Great Basin, Lake Isabella Basin, and Tehachapi Mountains bearing on the issues of population movements, changes in land-use patterns, and subsistence-settlement organization (Basgall and McGuire 1988; Bettinger 1976, 1977; Bettinger and Baumhoff 1982; Cuevas 2002; Delacorte 1999; Delacorte et al. 1995; Delacorte and McGuire 1993; Dillon 1988; Gilreath and Hildebrandt 1997; Harrington 1957; Hildebrandt and Ruby 2000; Lanning 1963; McGuire et al. 1982; Reynolds 1996; Schroth 1994; Sutton 1994; Yohe 1992; and others).

Uniqueness of Area for Testing Numic, Tubatulabalic, and Earlier Prehistory

The study area is uniquely suited to evaluations of linguistic and archaeological models related to the events and timing of Northern Uto-Aztecan prehistory. Circumstances in the study area permit testing of alternative models of Numic expansion (Aikens and Witherspoon 1986; Bettinger and Baumhoff 1982; Sutton 1987, 1991), as compared to the in-place development of the Tubatulabal linguistic pattern (see below). Archaeological data might be expected to mirror strikingly dissimilar consequences if the two areas experienced contrasting cultural developments (Numic inmigration and population replacement versus in place development of the Tubatulabal).

Archaeologists and linguists agree that the Tubatulabal language is longstanding with 2,000–3,000 years of in-place development (Bettinger 1994, 2002; Foster 1996; Fowler 1972; Lamb 1958; Miller 1986; Moratto 1984). If that is the case, archaeological evidence would be expected to verify the initial Tubatulabal colonization at that early date with a subsequent unbroken record of cultural development. There is little evidence of any earlier cultural stratum in the Tubatulabal homeland. The Tubatulabal probably migrated into an uninhabited territory or one only minimally occupied by competing populations. This pattern, different from that related to the Numic and pre-Numic region, may be detected and compared.

Territories of Historical Ethnolinguistic Groups

The research area is a mosaic of contiguous ethnic territories (Figure 1.11). An ethnographic boundary cuts across the research area, running along the crest of the Sierra Nevada, separating the Numic and

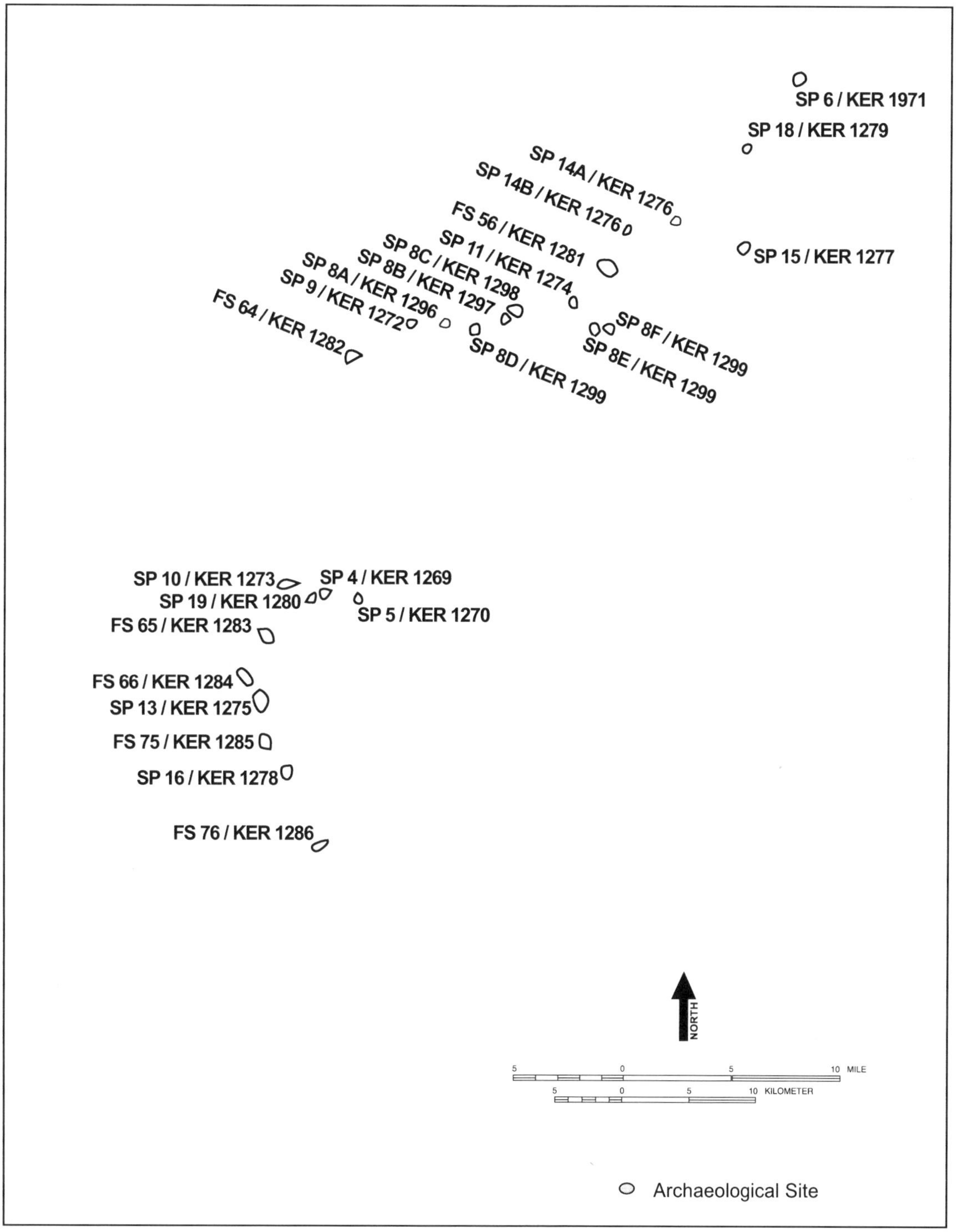

Figure 1.5 Scodie Mountain Sites.

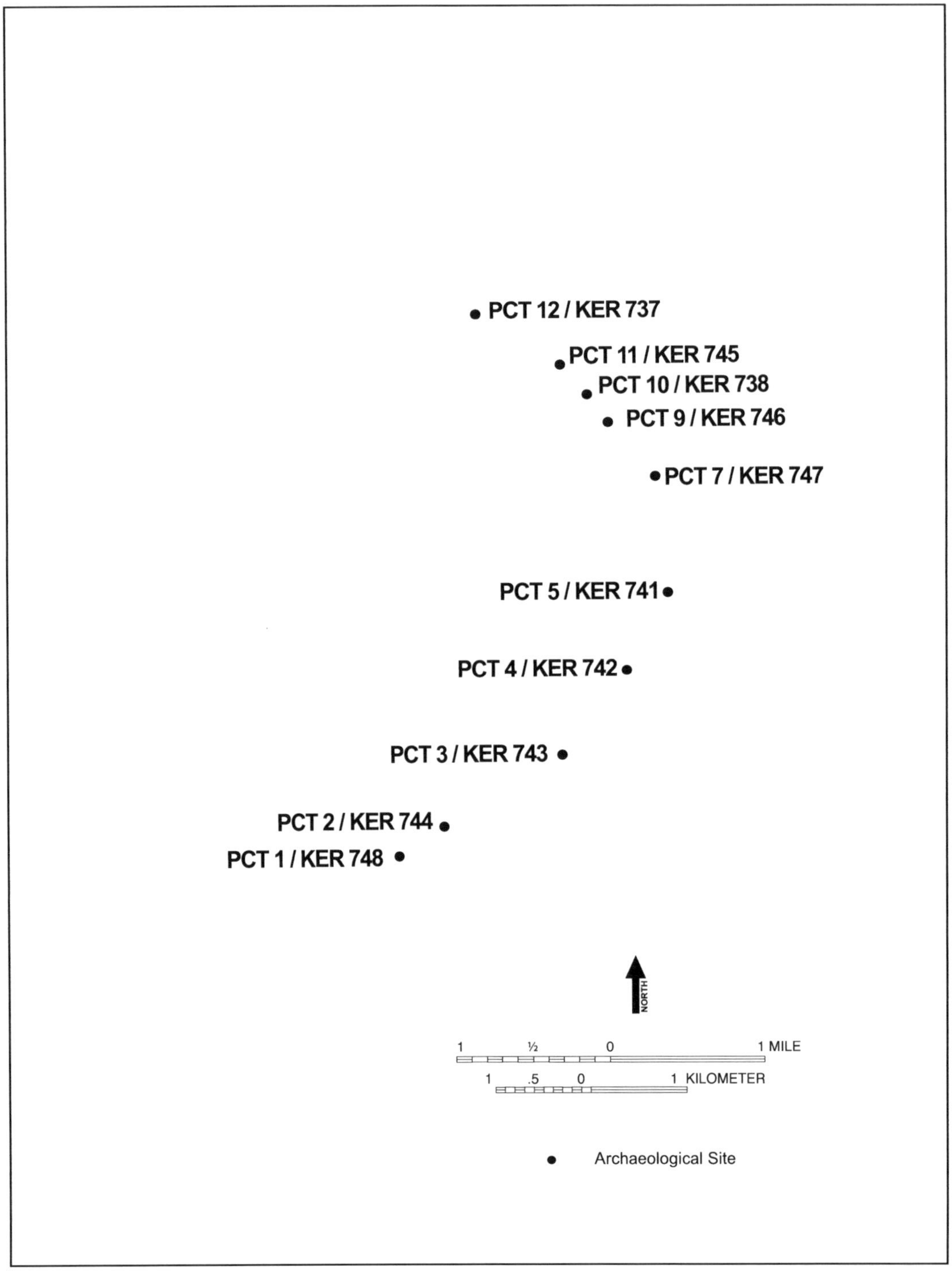

Figure 1.6 Morris Peak Sites.

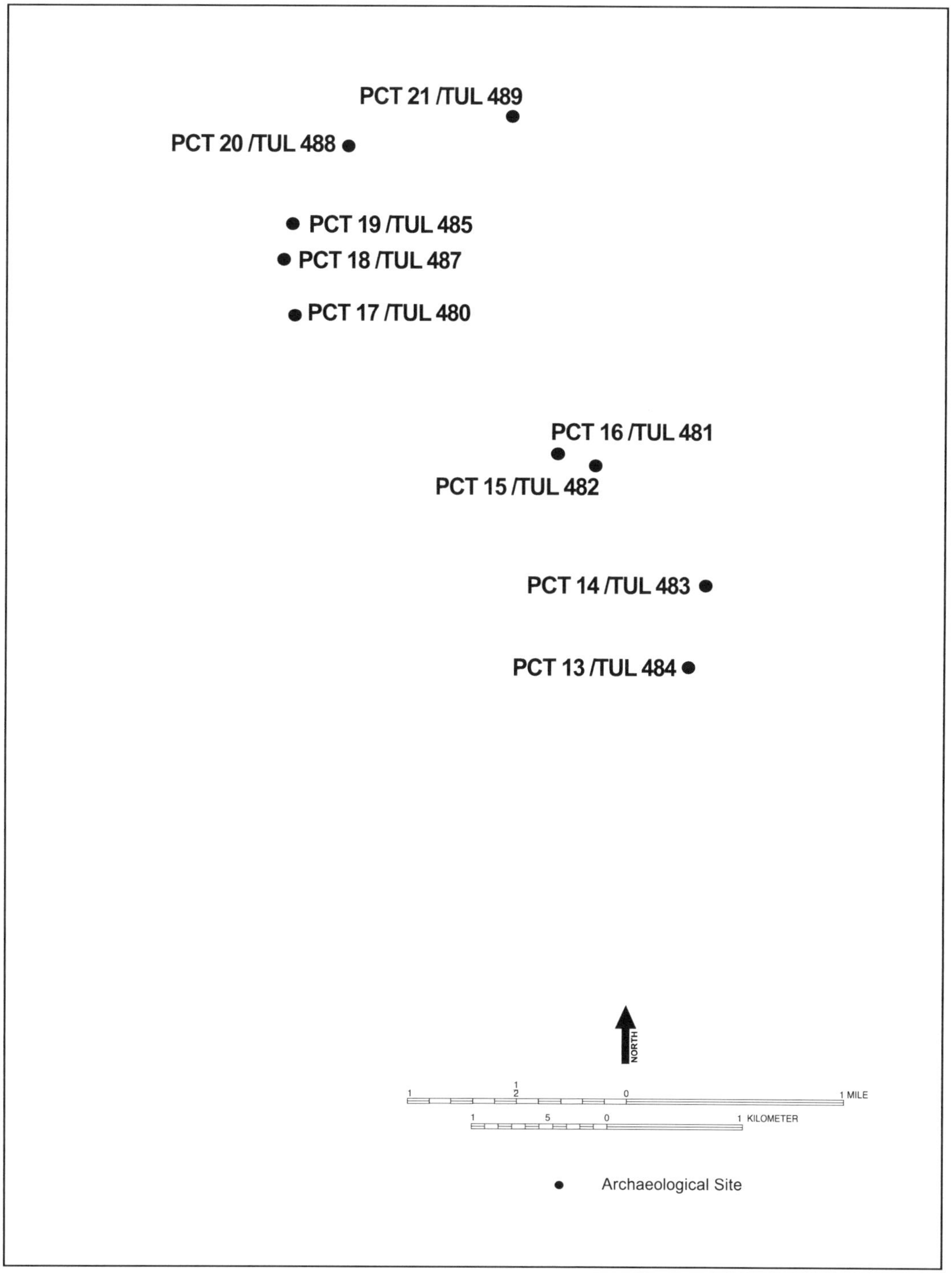

Figure 1.7 Lamont Meadow Sites.

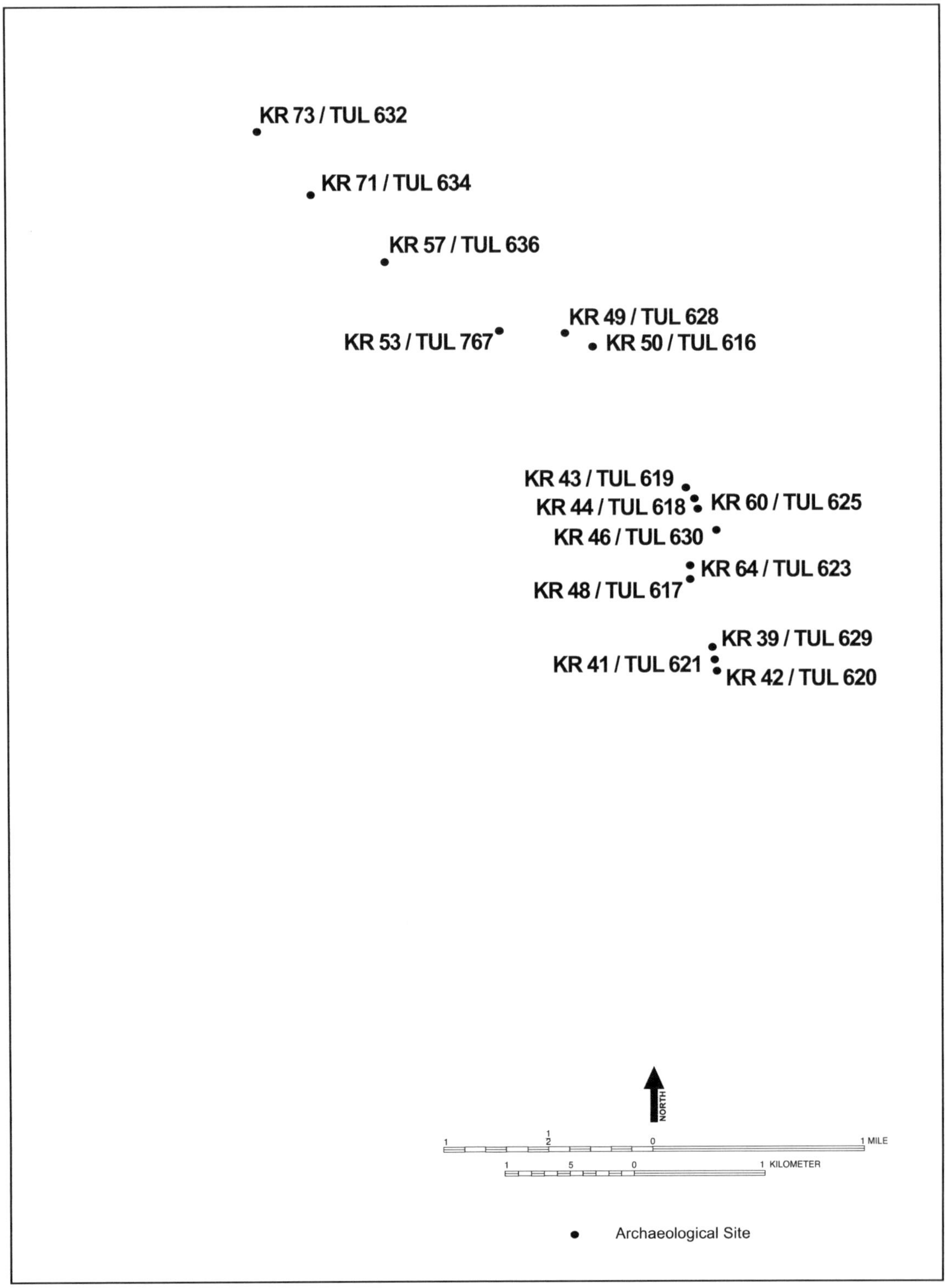

Figure 1.8 Bear Mountain Sites.

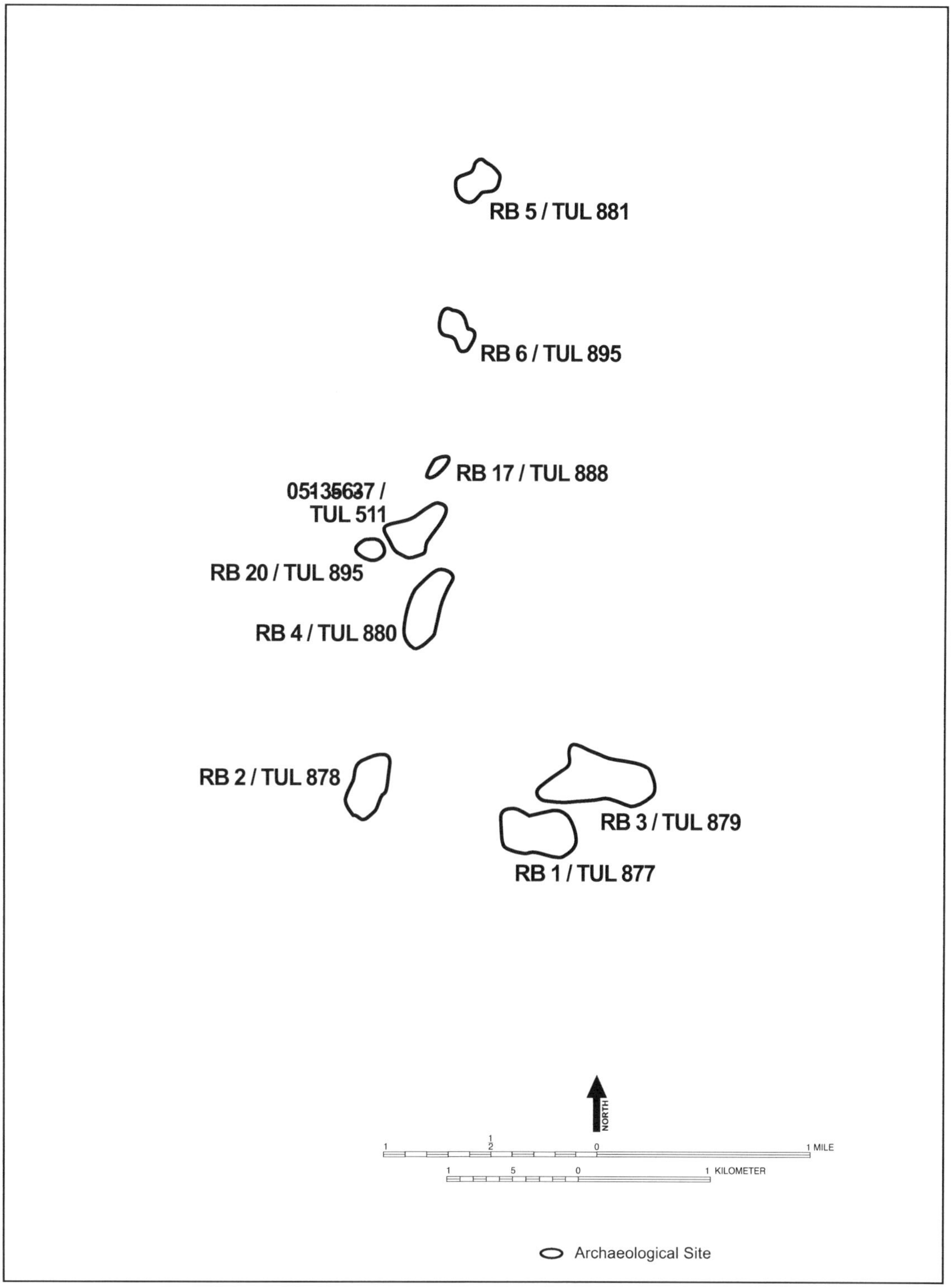

Figure 1.9 Rockhouse Basin Sites.

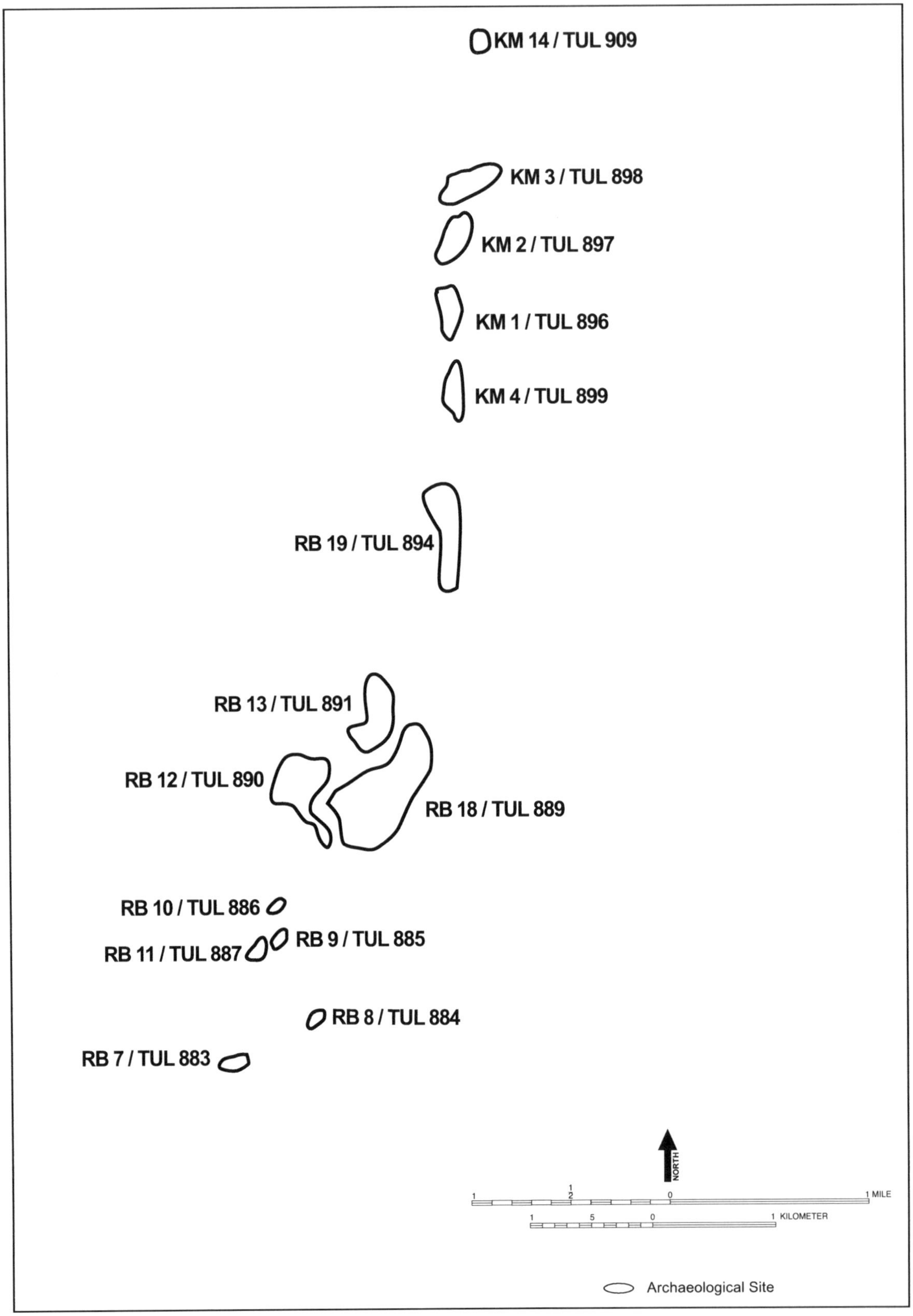

Figure 1.10 Rockhouse Basin and Kennedy Meadow Sites.

Tubatulabalic homelands (Driver 1937; Grosscup 1977; Steward 1938; Voegelin 1938; Zigmond 1986). Besides this east-west division, a north-south boundary separates the Tubatulabal from the Panamint Shoshone to the north and the Kawaiisu to the south.

Numic groups, the Panamint Shoshone and Kawaiisu, were historically confined to the relatively arid parts of the Tehachapi Mountains and southwestern Great Basin. The region west of the Sierra crest and the drainage of the South Fork of the Kern River was home to their neighbors, the Tubatulabal, who are linguistically and culturally distinct and had more abundant resources (Driver 1937; Grosscup 1977; Steward 1938; Voegelin 1938; Zigmond 1986).

Sociopolitical and residential organizations varied between these native societies. Highly formalized social structures and a relatively centralized village system typified the more populous Tubatulabal (Kroeber 1925; Voegelin 1938), while highly mobile family bands characterized the Kawaiisu and Panamint Shoshone (Steward 1938; Zigmond 1986). These distinctive populations may have left archaeological markers of their sociopolitical and ideological differences. Each group occupied different core territories and differed in land-use patterns and seasonal movements. Therefore, it may be possible to reconstruct territorial or social patterns from the archaeological record and date those signatures. Such reconstructions might serve to mark the territories of the language groups and help delineate population shifts or in-place developments over time.

Research Objectives

This study evaluates competing models of prehistoric population and language movement versus *in situ* cultural change in the far southern Sierra Nevada and southwestern Great Basin. In simple language: either the Numic expansion took place or it didn't (Hughes 1994:68–69). I hope to clarify and critically examine the validity and implications of these models. As well archaeological evidence is reviewed to support the most succinct and compelling model.

Much of the problem with evaluating population movements is that competing hypotheses are not fully developed and tested. Thus the conclusions are more or less ambiguous. Population movements have often been inferred from simple changes in the archaeological record involving only a few artifact types or a solitary change in subsistence patterns. Such inferences are inherently weak. Similar evidence could also be used to support *in-situ* cultural reorientations (e.g., cultural adjustments to environmental changes with no ethnolinguistic replacement).

What is required is an approach in which competing hypotheses are compared and data are evaluated to support or refute the alternatives. This study tests the null hypothesis that population movements *did not* take place rather than merely providing supporting evidence that movements did occur. To reduce the discussion to the most basic elements, two sets of alternatives exist — replacement models and in-place models of development.

Replacement Models

Historical linguists and some archaeologists paint a picture of changing linguistic distributions over time and space, suggesting various population movements and replacements in the study area. If the scenarios of historical linguistics are accurate, the archaeological record of the last 5,000 years of Northern Uto-Aztecan prehistory might include evidence for: (1) the arrival of Northern Uto-Aztecan and pre-Numic groups; (2) the divergence of Northern Uto-Aztecan into its several language families; (3) the divergence of Numic from Tubatulabal; (4) the differentiation of Numic into its several branches; and (5) a Numic

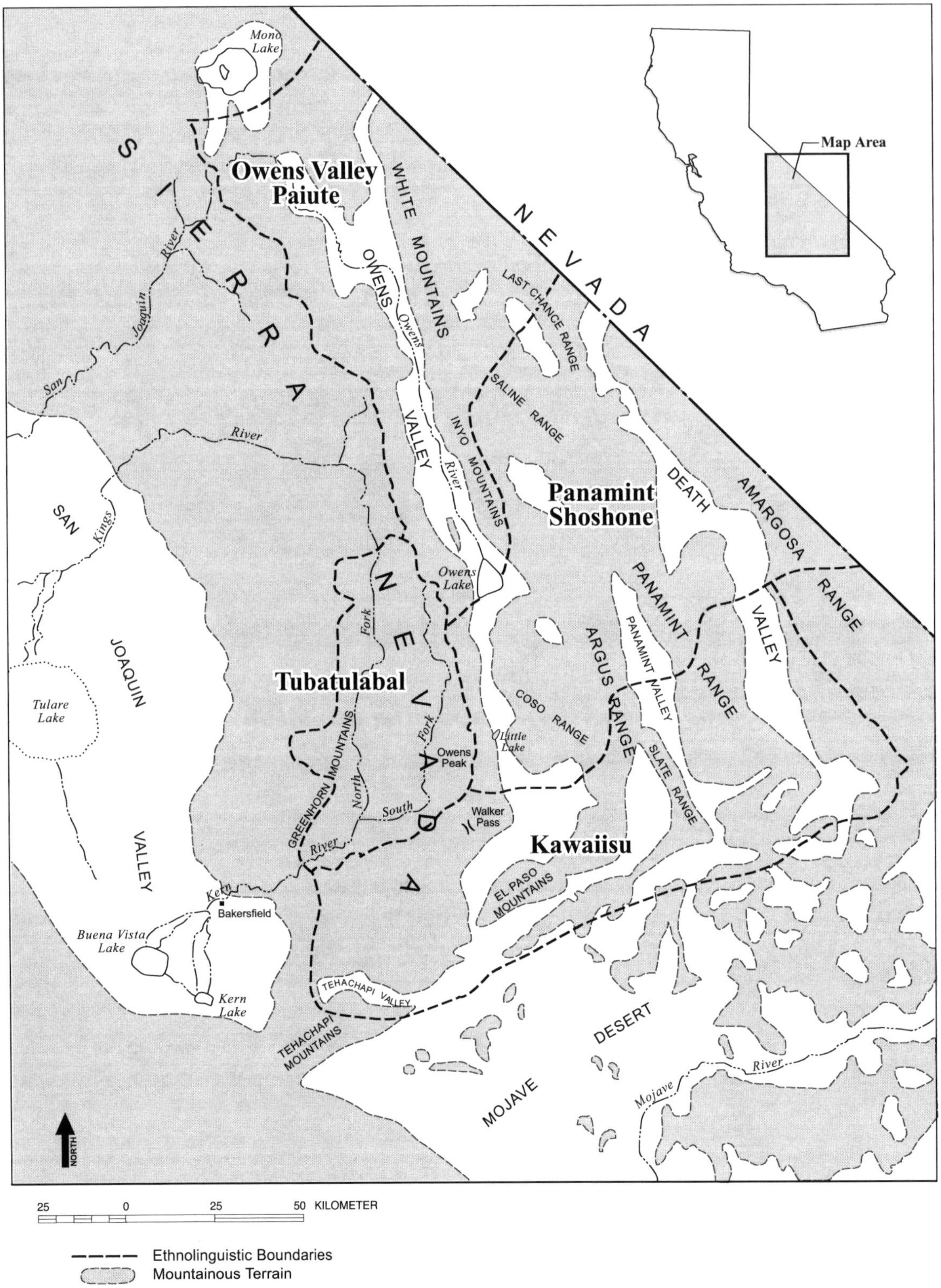

Figure 1.11 Ethnolinguistic Divisions of Eastern California.

expansion. Few speculations can be made regarding the arrival of Northern Uto-Aztecans, as data are scant from this early time period. However, the arrival of the Tubatulabal, the initial colonization and influx by pre-Numic peoples, the divergence of the Tubatulabal from the Numic, the replacement of pre-Numic groups by Numic populations, and the expansion of Numic colonists into the Great Basin can be evaluated using archaeological materials from the study area.

Models of In-Place Development

Alternative scenarios, based on historical linguistics and archaeology, provide varying perspectives on the direction and timing of Northern Uto-Aztecan prehistory and, more specifically, Numic and pre-Numic prehistory (e.g. Aikens and Witherspoon 1986; Garfinkel 1982; Pearson 2002; Whitley 1998). If Numic is longstanding, as several researchers have argued, then the archaeological record in the study area might provide evidence for a relatively continuous unbroken cultural record of gradual, *in-situ* changes and continuity (Grant et al. 1968; Warren 1984:384; Warren and Crabtree 1986:192). Since the exact location of the proto-Numic homeland remains to be determined, and the study area lies within the larger area believed by many to have been that linguistic hearth, a lengthy in-place development is a distinct possibility.

Several lines of archaeological evidence support gradual change and continuity, rather than indicating abrupt change and population replacement (Warren 1984:384; Warren and Crabtree 1986). Rock art data have been central in interpretations arguing for ethnic continuity. Historic and prehistoric connections and continuous evidence for in-place Numic cultural development would support such a view (Grant et al.1968; Pearson 2002; Warren 1986:384; Whitley 1998:53–60).

The linguistic evidence can be seen as rather ambiguous. After reviewing the evidence, some linguists are persuaded that cultural stability could have been the case. They argue for continuity rather than linguistic change (Goss 1977; Shaul 1986). The linguistic diversity found in the southwestern Great Basin is explained as accretion rather than parentage (Golla 2000; Hill 2002; Nichols 1992).

My purpose here is to objectively evaluate the archaeological evidence and examine alternative models of prehistoric population development(s). I consider models that portray ethnic continuity and evaluate them with respect to whether in-place Numic development is consistent with the archaeological record.

Alternative Test Implications

The Direct Historical Approach: Examining known ethnographic (historic) village sites and subsistence areas should provide archaeological records of Numic and Tubatulabal activities. Tracing the archaeological patterns back in time from the historical period would clarify the antiquity and development course of these patterns.

Cultural Sequence and Regional Chronology: Population movements and displacements should be recognized as abrupt changes in the archaeological record of sites or localities. I suggest that the arrival of a language group will be visible archaeologically as a significant shift from a prior pattern and should not be associated with the relative stability or gradual changes characterized by the persistence of a culture (Basgall 1982). Therefore, breaks in the cultural sequence often imply population movements.

If the Numic/pre-Numic population replacement took place in the study area, then that should be reflected by distinctive changes in site histories mirrored in obsidian hydration chronologies. Widescale disruption in the regional patterning of site occupations would indicate population movements, immigration, expansion, displacements, and cultural discontinuities. Widespread regional patterns, indicating a disruption or hiatus in occupation, at the same time as hypothesized population influxes or migrations, would provide credible evidence supporting cultural discontinuity.

Land-use shifts, if identified, should be coincident with the proposed timing and geographic placement of hypothesized population displacements. Also, differences in settlement-subsistence strategies might be a function of population shifts (Bettinger and Baumhoff 1982).

Subsistence-Settlement Regimens: Population movements often co-occur with adaptive shifts, and such shifts must be explained with reference to some advantage for the successor population. Changes in adaptive strategies and the introduction of the intensification of resource use might signal a new population and language. Adaptive change, however, could also reflect *in-situ* intensification. Distinguishing between the two is relatively difficult.

If subsistence-settlement patterns change, then two patterns (for both Numic and pre-Numic groups) might be exhibited simultaneously in the proto-Numic homeland region (Hughes 1994; Madsen 1994). If contemporary but different adaptive strategies occur, then these expressions could still represent a, single ethnolinguistic group's seasonal round. To lend less support to this possibility, the subsistence patterns should be largely incompatible. Some subsistence activities might be precluded because of scheduling conflicts or because of the incompatibility of the targeted resources, because of their location and distribution. Many factors could prevent the incorporation of two disparate food procurement strategies. Several scholars (Hildebrandt and Ruby 2000; Madsen 1986; Pippin 1977) believe that it would be difficult to conduct both upland, green-cone, piñon harvests and lowland, communal bighorn sheep hunts. Ethnic signatures for the separate populations would support the position that two cultural groups occupied the same area simultaneously. A compelling argument can be made if two distinctive subsistence-settlement systems occupying similar environments but using different aspects of the resource base could be documented.

Site counts within a defined area might increase during times of population increase. Alternatively, site size and character might change as a function of larger numbers. Such changes could result from immigration, especially with reference to conservative hunter-foragers. Alternatively, such changes could also be caused by technological innovations that would allow for expanded, *in-situ* population growth.

Toolstone Access, Exchange Relations, and Territoriality: Studies of Coso obsidian use document changing production patterns and suggest shifts in resource use over time (Gilreath and Hildebrandt 1997:179). Further, regional settlement-subsistence models depict reduced access to the obsidian source, decreasing mobility, and increasing territoriality over time. The time when Numic groups may have replaced pre-Numic populations in the area (ca. A.D. 600–1000) is also when territoriality was at its zenith and access to the Coso obsidian source appears to have been the most limited.

Ericson and Meighan (1984:150) remarked that changes at the Coso source itself might be principally responsible for the abrupt decline and cessation of trans-Sierran obsidian exchange, after A.D. 1000. Since the Coso source was supposedly once in the territory of the Tubatulabal (Steward 1938:x), a recent intrusion (after A.D.1000) of Numic people may have altered that direct and free access (Farmer 1937).

The relative abundance and spatial patterning of different toolstone materials at various archaeological sites may relate to this pattern of decreased accessibility. Sites may exhibit variable frequencies of flaked-stone materials (obsidian and cryptocrystalline stone). These differences in part may be functions of distance, but they also may indicate changing social relations and exchange patterns coincident with ethnic/linguistic boundary shifts, territoriality, and population movements (Bettinger 1982).

Numic and Pre-Numic Material Culture — Continuity and Discontinuity: The use of certain artifact types or series may end abruptly rather than gradually declining as another form becomes more common. Such an abrupt decline suggests cultural discontinuity. When cultural discontinuities are evident, use of longstanding artifact forms may cease abruptly, rather than trailing off in a gradually diminishing pattern more typical of incremental, *in-situ* change. Gradual transition versus abrupt shifts in the occurrence of artifact types could equate to *in situ* cultural continuity versus discontinuity.

Symbols, signs, and beliefs are methods by which groups identify and distinguish themselves (Weissner 1983). Definitions of an ethnic group often incorporate an ideological element. Few elements of the archaeological record directly relate to the ideological dimension or religious realm; burial treatments and regional rock art styles might be among them. These aspects of the archaeological record would be sensitive indicators of cultural continuity (or discontinuity). The clustering of certain patterns of burial treatment or styles of rock art would be critical elements of material culture that do not easily pass through group boundaries. Rock art styles have often been used as proxy indicators of cultural traditions (linguistic or ethnic groups). I would argue that they are the most direct linkage within an archaeological assemblage for reconstructing former ceremonial systems and cultural groups (Garfinkel 1982; Lee and Hyder 1991). Rock art as stationary, iconographic monuments (not subject to the vagaries of exchange) may serve explicitly or implicitly as demarcations between groups.

In the study area, several researchers have argued for strong continuities in the design traditions exhibited in the Coso Style petroglyphs and pictographs (Garfinkel 1982; Grant et al 1968; Pearson 2002; Warren 1984; Warren and Crabtree 1972; Whitley 1998). These researchers see a continuous, unbroken cultural tradition from circa 1000 B.C. or earlier through the historic era. Recently, through the use of obsidian hydration measurements and studies of archaeological contexts, more refined dating methods are now being applied to the Coso rock art traditions (Gilreath 1999, 2003).

Another study has also reviewed available evidence for the form and dating of prehistoric burial patterns in the region (Gilreath 2000). This information may be used to address the question of continuity versus discontinuity. Dating of these unique rock art styles might allow them to be identified with either Numic or pre-Numic inhabitants of the region and help assess the stability or interruption of cultural patterns.

Summary

Given certain limitations, linguistic archaeology can illuminate the past, serving as a valuable source for generation of hypotheses about prehistory. Artifact styles may be especially sensitive indicators of ethnic groups and their boundaries. When several stylistic elements co-occur spatially and temporally, ethnic groups and their boundaries may be indicated. Material cultural elements relating to the ideological and religious realms (e.g., burial patterns and rock art) may be some of the most germane elements bearing on ethnicity.

Northern Uto-Aztecan linguistic groups are recognized historically from the study area. Tubatulabal and Numic peoples have occupied the region for some time. Tubatulabal is thought to have a long, in-place development of 2,000–3,000 years. Numic peoples may either be relative newcomers or have a long in-place development. Contrasting models of population movements or cultural continuity may be evaluated. Using the direct historical approach, Tubatulabal, Kawaiisu, and Panamint Shoshone affiliations for historical archaeological sites may be identified and the antiquity of such patterns defined.

Immigration and population displacement might correlate with abrupt changes in the archaeological record: the hinge points for different chronological periods or discontinuities in regional and site-specific obsidian hydration measurement curves. If simultaneous divergent subsistence-settlement patterns are documented, then these patterns might represent distinctive elements of the seasonal round for multiple ethnolinguistic groups. Toolstone use may reflect changing access patterns, exchange relations, and territoriality. All figure prominently into the shifting land-use patterns expected when population movements take place. Spatial patterning and dating of regional rock art styles and burial treatments can help clarify whether a continuous, unbroken cultural tradition existed from early prehistoric times (ca. 3000 B.P. or earlier) through the historic era or whether one or more episodes of population replacement occurred.

Chapter 2
Environmental Background

Scope and Purpose

The relationship between prehistoric foragers and their physical environments has long been a research interest of anthropologists (Antevs 1952; Steward 1938). Many aspects of the environment profoundly influenced the behavior of aboriginal hunters and gatherers. Anthropological work in the realm of cultural ecology predisposes us to think of the influence of animal and plant distributions on foraging and mobility strategies (Steward 1933, 1938, 1941). Yet the relationship between environment and culture is a frequent and sometimes controversial topic in California and the Great Basin (Madsen 1981; Zeanah 2002). Environmental factors interacted in complex ways to affect the settlement and subsistence decisions of the native peoples whose territories crosscut the study area. This chapter provides an environmental context for the discussions that follow. Included are brief reviews of local geography, geology, geomorphology, soils, and climate.

Steward (1938:27–28, 232) stressed the importance of piñon nuts as a storable staple, the abundance of which determined, to some extent, the location, size, and permanence of winter villages for aboriginal peoples throughout the Great Basin. Various scholars have applied the general principles of behavioral ecology and diet breadth models to consider the costs and benefits of piñon use (Bettinger 1976, 1977; Simms 1985; Wells 1983). Knowledge of the expansion and contraction of piñon resources throughout prehistory could help us to predict how and when prehistoric hunter-gatherers might have modified their subsistence-settlement strategies in response to changes in resource availability (Rhode and Madsen 1998; Zeanah 2002). Accordingly, this chapter summarizes regional paleoclimatic changes including the abundance and distribution of local piñon woodlands.

Pre-agricultural foragers practiced seasonal movements, timing their settlement decisions to the availability of key plant and animal resources. For a better understanding of such decisions, an overview of local flora and fauna is also included. Additionally, since stone tool material types in the archaeological record can reflect territorial extent, cultural preference, trade, technological variables and even temporal variability, toolstones' sources and locations need to be considered. The chapter closes with a brief treatment of the rocks used in the manufacture of flaked and groundstone tools.

Study Area and Sites: Character And Location

The 96 prehistoric site components examined are situated along the route of the Pacific Crest Trail in the southern Sierra Nevada within the Kern Plateau and Scodie Mountains (Figure 1.3). The sites lie along a 35-mile (60 kilometer) segment of trail running from the Scodie Mountains, across Walker Pass, traversing Morris Peak, Lamont Meadow, Chimney Meadow, Bear Mountain, and Rockhouse Basin, and terminating in the north at Kennedy Meadow (Figures 1.3, 1.5–1.10). This crestal zone forms a sharp environmental boundary, separating the relatively well-watered Kern River drainage from the semi-arid regions of the Great Basin and the Mojave Desert.

The Kern Plateau region, composed of granitic domes and ridges, rises in elevation from 5,000 to 8,000 feet (1700–2600 m.). Two natural passes were used by native peoples: Walker Pass and Dove Springs Pass (Figures 1.2 and 1.3). The Scodie Mountains are a northern extension of the Tehachapi Mountains. The Tehachapis are a series of northeasterly trending ridges that begin near Keene and end in the tortuous terrain of upthrust volcanics in the Cache Peak locality (Figure 1.2). Long thought to be an extension of the Sierra, the Tehachapis are now recognized as an independent, uplifted, and faulted mountain block (Hill 2005).

Kern and South Fork Valleys

The Kern and South Fork Valleys separate the Greenhorn Range, the Kern Plateau, and the Piute Mountains. The Kern River flows southward through the Kern Valley and into Lake Isabella and drains the Kern Plateau. The river's source is in upper Kern Canyon, the main drainage for the western and central part of the range. The misnamed South Fork Valley is where the Kern River flows westerly. The Kern River makes an abrupt turn in the South Fork Valley, changing direction from its predominantly northward flow (Figure 1.2).

Mojave and Great Basin Deserts

To the east is the interface between the Great Basin and the Mojave Desert. The Great Basin is characterized by north-south-trending fault block ranges (horsts) and intervening basins (grabens). Owens Valley, just northeast of the study region, is one element in this system. The Mojave Desert consists of broad plains with sinks and "buttes"; the Rand Mountains, El Paso Mountains and Indian Wells Valley are geographically part of this area (Hill 2005). The numerous "buttes" (misnamed) are actually the cores of ancient Tertiary volcanoes.

Geology, Geomorphology, and Soils

The Sierra Nevada is an enormous batholith — an immense, predominantly westerly tilted, fault block of granitic rock (Hill 2005). The Sierra's western face is dissected by a series of rivers that flow into California's San Joaquin Valley. The Kern River is the southernmost of these. The extreme southern Sierra is a region dissected by erosion and characterized by roughly conical summits separated by long, level saddles and by canyons containing small, flat-bottomed meadows (Hill 2005).

The Sierra Crest escarpment is a deeply furrowed and precipitous wall of simple structure produced by erosion of a steep fault scarp. The ridgeline itself is formed by mechanical weathering, leaving a notable accumulation of unstable natural rock debris and rubble. Soils are poorly developed with little organic content and mainly composed of disintegrated granodiorite. The South Fork Valley itself has a broad floor of deep alluvial soils (Hill 2005).

Present Climate

The climate of the study area is influenced by the semipermanent Pacific high-pressure cell located over a large portion of the Pacific Ocean (Barbour and Billings 2000; Barbour and Major 1977, 1988; Barbour et al. 2006). Clockwise airflow movement out of this zone results in prevailing westerly winds. During the winter months this high periodically breaks down and is largely replaced by the southern extension of the Aleutian low-pressure cell causing storm conditions, snowfall, and variable winds. Increases in temperature during summer months create thunder and lightning storms of mostly short duration (Barbour and Billings 2000; Barbour and Major 1977, 1988; Barbour et al. 2006).

Summer temperatures range from maximums in the desert region of 80–105° F to minimums in the high mountains of 15–37° F (40 to -10° C). Winter temperatures vary widely from highs of 55–70° F (22 to -29° C) in the desert region to lows of 0 to -20° F in the higher mountains (USDI 2001).

As is apparent from these temperature ranges, winters can be extremely cold in the high elevation regions where there is accumulated snowfall. Precipitation comes in the form of both rain and snow, more than half falling from January to March. The higher elevations within the southern Sierra receive up to 38 inches (960 mm) of precipitation annually (USDI 2001).

Paleoclimate and Prehistoric Dynamics of Piñon Woodland

Native American lifeways are closely linked to the location and fluctuating abundance of key subsistence resources. The appearance and disappearance of basic food resources would have been very important for aboriginal peoples. In the study area, the most important resource for ethnographic groups was piñon nuts. The presence or absence and extent of this single resource can be tracked through time using a variety of proxy indicators including woodrat middens and pollen cores.

The following overview is based on a variety of recent syntheses, especially Halford (1998) and Wigand (2002). The primary data for this synthesis include information from vegetation histories and proxy climatic indicators provided by Graumlich (1993), Graybill et al. (1994), LaMarche (1973, 1974), Mehringer (1977, 1986), Stine (1990, 1995), Thompson (1990), Van Devender and Spaulding (1979), and Wigand et al. (1995).

The expansion of piñon pines into the Great Basin is correlated with the disappearance of the treeless sagebrush steppe and the end of the extended cool conditions of the glacial period. Piñon pine appeared in the White Mountains about 12,000 B.C. (Jennings and Elliot-Fisk 1993) (see Table 2.1 for details of radiocarbon dates and calibrated ages). This expansion appears to have been quite rapid in response to the warm and moist conditions characterizing this period (Wigand 2002). Other data also suggest that piñon pine was also spreading upward in elevation (Wigand et al. 1995).

The Middle Holocene (6000 to 3500 B.C.) saw a retreat of the piñon pines to higher elevations as a result of intense regional drought. This period is further characterized by a disappearance of piñon in the paleoclimatic record from lower elevations. But contrary to earlier generalized paleoclimatic reconstructions, considerable local and regional variability seems to have accompanied the middle Holocene thermal maximum, with one or more brief episodes of cooler, wetter climate occurring around 4500 to 2300 B.C. (Wigand et al. 1995). Halford (1998) documents the reappearance of piñon pines in the Bodie Hills at 3300–4000 B.C., and similar dates coincide with piñon's reappearance in the Inyo and White Mountains (Reynolds 1996) (See Table 2.1).

Following the onset of wetter conditions ca. 3500 B.C., piñon began to reappear at lower elevations on peaks higher than 7,500 feet (2300 meters). Near the study area, peat occurred at the base of the Little Lake pollen record, ca. 3000 B.C., signaling the rejuvenation of springs in the region (Mehringer and Sheppard 1978). During the neopluvial period from 2000 B.C. to A.D. 1, piñon pine again expanded into the Sierra Nevada and southwestern Great Basin in association with winter-dominated rainfall (Eerkens and King 2002). During this period, piñon pine was found at lower elevations where it had not occurred before (Wigand 2002).

Table 2.1 Radiocarbon and Calibrated Dates for Sources of Paleoclimatic Data.

Laboratory Number	Uncorrected Radiocarbon Age (yr B.P.)	Uncorrected Radiocarbon Date	Uncorrected Range of Radiocarbon Date at 2 Sigmas	Calibrated* Range of Radiocarbon Date at 2 Sigmas	Locality/ Site Name	Sample Substance	Reference
Unavailable	8790 ± 110	6840 B.C.	6730–6950 B.C.	cal 7600–8030 B.C.	White Mountains, Falls Canyon 1	Woodrat Midden Piñon	Jennings and Elliot-Fisk (1993)
Unavailable	4980 ± 80	3030 B.C.	2870–3190 B.C.	cal 3645–3950 B.C.	Bodie Hills DLP150896PEW1	Woodrat Midden Piñon	Halford (1998)
WSU1464	5050 ± 140	3100 B.C.	2820–3380 B.C.	cal 3628–4115 B.C.	Little Lake	Lake sediments	Mehringer and Sheppard (1978)
Beta 86097	4690 ± 60	2740 B.C.	2680–2800 B.C.	cal 3361–3541 B.C.	Inyo Mts, Papoose Flat	Woodrat Midden Piñon Seeds	Reynolds (1996)

KEY: *cf. Stuiver and Reimer 1993; Stuiver et al. 1998

Over the last 2000 years, piñon pine has significantly expanded (Wigand 1997). The largest expansion was about 1400–1000 years ago, when there was a shift to more summer rainfall and slightly cooler climate that encouraged seedling establishment.

Piñon expansion was primarily a regional phenomenon during the Late Holocene; a period of retrenchment is documented from ca. AD 1000 to 1350, as shown by the pollen record and woodrat middens, and this dramatic climatic episode even affected higher elevations. This "Medieval Warm Epoch" (Medieval Climatic Anomaly) is verified by tree ring data and relict tree stumps below the current surfaces of Lake Tahoe and Mono Lake (Stine 1990, 1995). Stine (1990), in fact, sees several extended droughts terminating at about A.D. 1150, 1350 and 1600 and followed by brief wet events.

From AD 1600–1650, piñon woodland again expanded, corresponding with stronger winter precipitation and cooler temperatures. At approximately A.D. 1800 the piñon-juniper woodland increased to about two and half times its previous distribution in response to increases in mean annual temperature (Wigand 2002).

Probably the most significant shift in climate that could have affected the aboriginal use of the study area was the shift to cool-moist climatic conditions at 2000 B.C., with the onset of the Medithermal (Antevs 1952). Moratto et al. (1978) have similarly suggested that climate-based environmental changes at this time substantially affected human populations in the southern Sierra Nevada foothills and may have caused a rapid increase in upland occupation. Expansion of the lower margins of the piñon areas in the southern Sierra Nevada at about this time might have ultimately led to a new subsistence regime dependent on the use of piñon resources (McGuire and Garfinkel 1980).

The drought conditions documented during the period A.D.1000–1350 may have also led to environmental deterioration. Desiccation, especially in the adjacent lowland desert environments, could have fostered more intensive use of upland resources, abandonment of some lowland localities, and perhaps prehistoric population movements (Bettinger 1991; Bettinger and Baumhoff 1982; LaMarche 1973).

Specific Vegetation Patterns

The far southern Sierra is a region of remarkable environmental diversity. There is great vertical relief with abundant water in some parts and aridity in others. See Figures 2.1–2.4 for examples. Because of this environmental diversity, a useful way to classify the region is by vegetation series (now also being called alliances). The Sawyer and Keeler-Wolf (1995, 2006) classification system includes a Montane Meadow Series found at high elevations in riparian associations (Rockhouse Basin) and well-watered flats (Chimney Creek, Lamont Meadow, Kennedy Meadow), a Single Leaf Piñon-Utah Juniper Series in the mountainous areas (Morris Peak, Bear Mountain, Scodie Mountain, Tehachapi Mountains), and a Joshua Tree Series (Indian Wells Canyon, Walker Pass, and Kelso Valley down to the South Fork Valley to Lake Isabella) in upland canyons and valley systems. All of these series are described in greater detail by Sawyer and Keeler-Wolf (1995).

Single-Leaf Piñon-Utah Juniper Series

A homogenous, dense growth of piñon and juniper blankets the steeply sloped mountain faces and ridgelines of much of the study area. Piñon pine (*Pinus monophylla*) is by far the most dominant conifer and the most abundant arboreal species (Hickman 1993:120). Density of these diminutive pine trees ranges from 2 to 15 individuals per 100 m^2 with generally from 5 to 40% ground cover. Two other pines occur within the study area; grey pine (*Pinus sabiniana*) at the lower elevations and Jeffrey pine (*Pinus*

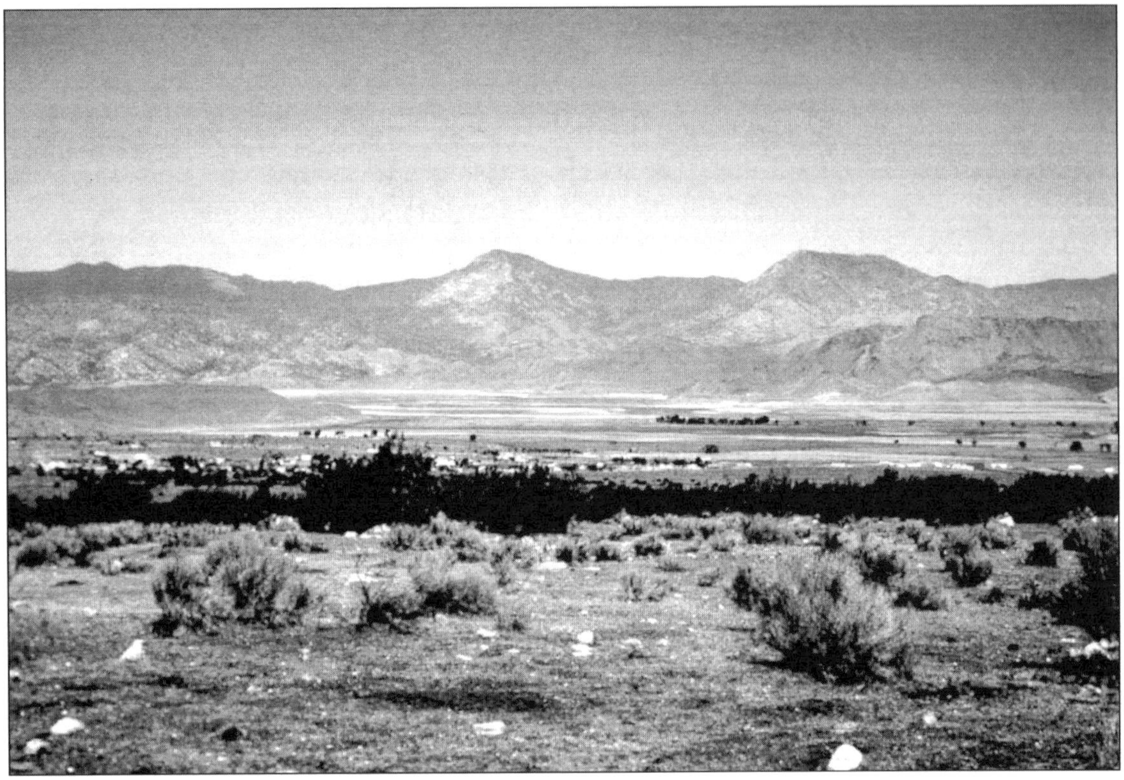

Figure 2.1 Isabella Basin and South Fork Valley.

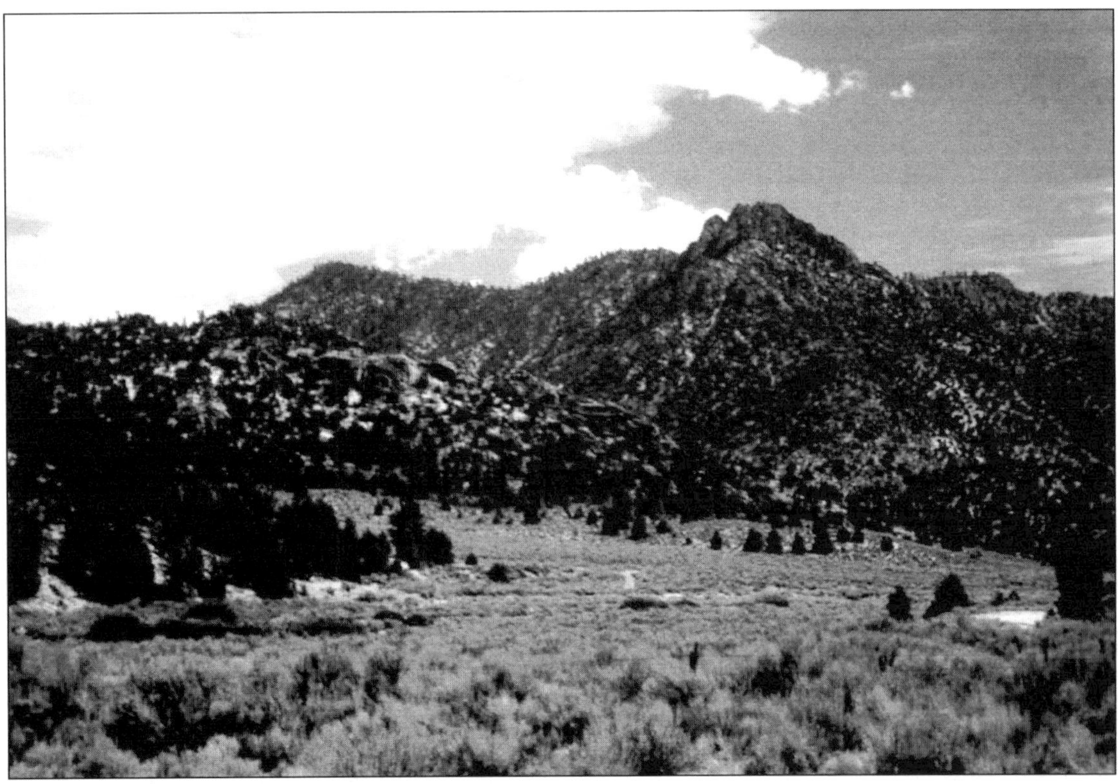

Figure 2.2 Lamont Meadow and Piñon Woodlands of the Kern Plateau.

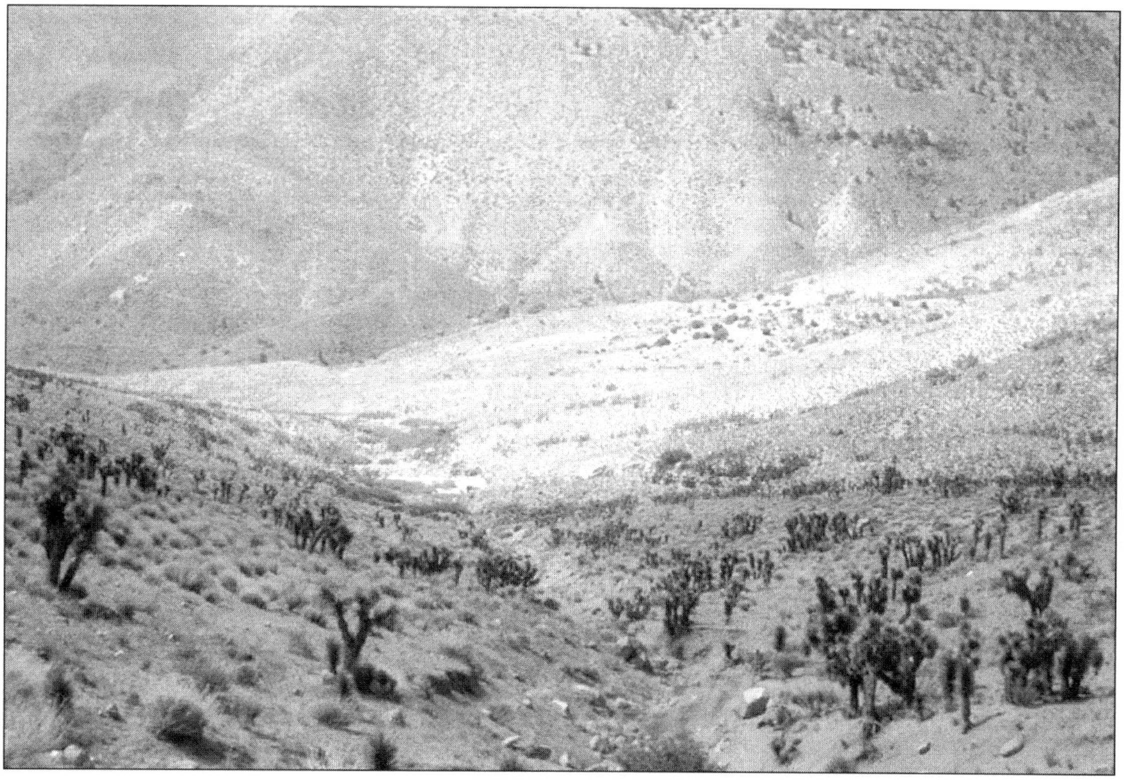

Figure 2.3 Joshua Trees and the Eastern Scarp of the Far Southern Sierra Nevada.

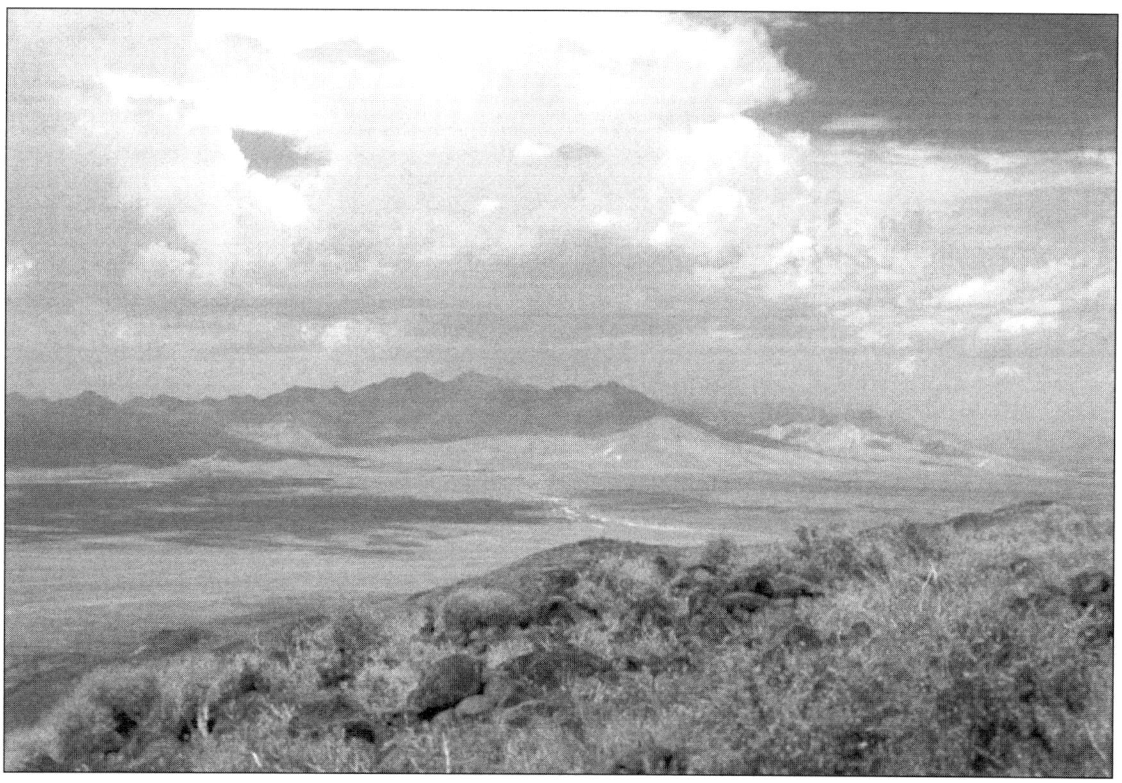

Figure 2.4. Indian Wells Valley and the Southern Sierra Nevada, Looking West From the Top of Black Mountain in the El Paso Range.

jeffreyi) at higher elevations, mostly in the northernmost area. Plant nomenclature follows Hickman (1993). Utah juniper (*Juniperus occidentalis*) and canyon live oak (*Quercus chrysolepis*) are usually found in rockier areas or steep slopes below ridges. Numerous understory shrubs, include sagebrush (*Artemisia tridentata*), rabbitbrush (*Chrysothamnus nausesous*), buck brush (*Ceonothus cuneatus*), flannel bush (*Fremontodendron californicum*), California coffeeberry (*Rhamnus california*), antelope brush (*Purshia tridentata*), creeping snowberry (*Symphoricarpos mollis*), and Mormon tea (*Ephedra viridis*). Herbaceous, low-growing plants also occupy this series and include desert needlegrass (*Achnatherum speciosum*), buckwheat (*Eriogonum* spp.), beavertail cactus (*Opuntia basilaris*), and bird's beak (*Cordylanthus rigidus*).

Montane Meadow Series

The Montane Meadow Series is found in small valleys and moist, well-watered flats associated with permanent water sources. This series also extends along the margins of the Kern River where it is associated with permanent streams, springs, shallow pools, and seeps. A great diversity of plants inhabits these localities. Dominant species include the ubiquitous willows (*Salix lemonii, S. gooddingii*), often growing in dense thickets; sedge (*Carex* spp.), rush (*Juncus* spp.), yerba mansa (*Anemopsis californica*), black oak (*Quercus kelloggii*), interior rose (*Rosa woodsii*), and various water-loving plants.

Joshua Tree Series

This plant series is found along the eastern scarp of the Sierra and in the vicinity of Walker Pass, South Fork Valley, and Kelso Valley. The most visible member of the lowland expression of this community is Joshua tree (*Yucca brevifolia*). Also present are western juniper (*Juniperus occidentalis)*, our Lord's candle (*Yucca whippleii*), sage (*Salvia dorii*), wild onion (*Allium campanulatum*), snowberry (*Symphoricarpos mollisi*), desert needlegrass (*Achnatherum speciosum*), sagebrush (*Artemisia tridentata*), flannel bush (*Fremontodendron californicum*), antelope brush (*Purshia tridentata*), bush lupine (*Lupinus albifrons*), and thistle (*Cirsium* spp.).

Food Plants

The most abundant and important food plants were harvested during late summer and autumn, principally in the Single-Leaf Piñon-Utah Juniper Series. Acorns from California black oak (*Quercus kelloggii*) and canyon live oak (*Q. chrysolepis*) were favored staples. Nuts from grey pine (*Pinus sabiniana*) and piñon pine (*Pinus monophylla*) were harvested during the fall. Considering the sheer bulk of the resource and the fact that piñon nuts were an ethnographically attested major winter staple, they must have been one of, if not, the most important economic plants for native populations to the area (Barras 1973; Butterbredt 1948; Cappannari 1950, 1960; Steward 1938; Voegelin 1938; Zigmond 1938; 1941). Few other plants are found in sufficient abundance within the vicinity of the study sites to support substantial aboriginal occupation (cf. Rhode 1980a, 1980b).

Animal Resources

Mule Deer

Voegelin (1938:11) and Smith (1978:444) record that deer were plentiful in the higher mountains of the far southern Sierra and were hunted almost year-round by the Tubatulabal. Deer were of lesser importance to the Kawaiisu and Panamint Shoshone (Steward 1938:80–83). Mule deer (*Odocoileus hemionus*) mostly occupy edges of forests and avoid dense vegetation, foraging on a wide variety of plants (Leopold 1951). Deer move in variously sized groups, with greatest dispersal and lowest concentrations during the summer (Longhurst et al. 1952). Many of the sites studied were probably occupied mainly during summer and fall. Individuals or small groups hunted deer at that time (Driver 1937; Steward 1938). In the event of early snows, herds of deer might have migrated downslope and could have been killed in larger numbers (Driver 1937; Heizer and Baumhoff 1962; Steward 1933, 1938).

Bighorn Sheep

Ethnographic data regarding sheep hunting by the Tubatulabal are notable by their absence (Smith 1978; Voegelin 1938). This omission suggests that bighorn sheep (*Ovis canadensis*) were uncommon within this region of the far southern Sierra (cf. Buechner 1960:11–16; Garfinkel et al. 1984:10–22). Peoples of desert areas to the east, where deer were scarce, focused instead on bighorn sheep as their most important large game animal.

Pictograph sites with images of bighorn sheep are found in the study area along the crest and eastern scarp of the Sierra, but these paintings are believed to have been created by people residing mainly in the desert areas to the east (Andrews 1980; Garfinkel 1978; Schiffman et al. 1982). Aboriginal inhabitants of the Coso Range, just east of the study sites, had a particular reverence for bighorn sheep which must have been the most important prey, judging by the thousands of petroglyph depictions found in the range (Grant et al. 1968; Heizer and Baumhoff 1962; Hildebrandt and McGuire 2003; Whitley 1998).

Pronghorn Antelope

The study sites are located outside the range of the pronghorn antelope (*Antilocapra americana*). Driver (1937) found no evidence of pronghorn drives within the territories of the Tubatulabal or Kawaiisu. Yet Steward (1938:82) documents communal drives for the Panamint Shoshone. Considerable populations of pronghorn inhabited the high desert plains just east of the Sierra Nevada in the time before Euroamerican settlement (Arkush 1995; McClean 1944).

The Tubatulabal and Kawaiisu participated in communal pronghorn drives outside their respective territories within the homeland of the Yokuts near Bakersfield in the southern San Joaquin Valley and with the Panamint Shoshone at what later became the community of Brown in the Indian Wells Valley (Kroeber 1925:528; Smith 1978:443; Steward 1938:13).

Rabbits

Two kinds of lagomorphs are found within the study area: hare or jackrabbit (*Lepus californicus*) and the cottontail (*Syvilagus audubonii*). Both were hunted, the former by drives and the latter with snares or bows and arrows (Driver 1937; Steward 1938:38–39, 80–82; Voegelin 1938:13). Jackrabbits are common to many biotic communities but are most abundant in the sagebrush scrub zone.

Rodents

Various rodents are found within the study area and, although they were not central to the aboriginal diet, rodents were taken when possible (Driver 1937; Steward 1938; Voegelin 1938). The western gray squirrel (*Sciurus griseus*), desert woodrat (*Neotoma lepida*), Mohave ground squirrel (*Citellus mohavensis*) and Botta pocket goper (*Thomomys bottae*) are animals recognized archaeologically within the study area. They were normally hunted using deadfall or rock traps. These animals were also run down, skewered, smoked, and flooded out of their burrows and killed with sticks and stones. They were normally only a minimal component of the Tubatulabal diet but were more central for groups living in the desert areas (Steward 1938; Voegelin 1938).

Fish

The South Fork and other tributaries of the Kern River provided excellent habitat for several native fish, including golden trout (*Salmo aguabonita*), rainbow trout (*Salmo gairdneri*), and Sacramento sucker (*Catostomus occidentalis*) (Moyle 2002). The Tubatulabal trapped, speared, and poisoned fish, and they gathered them in larger numbers during communal corralling (Voegelin 1938:14).

Birds

Birds (including waterfowl) are found within the study area but were never of great economic importance to the aboriginal peoples. Smith (1978:444) and Voegelin (1938:13) note that certain game birds were taken by bows and arrows or rock traps. These included goose (*Anser albifrons*), canvasback duck (*Aythye valisineria*), mountain quail (*Oreotyx pictus*), band-tailed pigeon (*Columba fasiata*), and blue-winged teal (*Anas discors*).

Toolstones

The El Paso Mountains contain outcrops of colorful cryptocrystalline silicate rocks, among them honey and blood-red jasper, rainbow agate, opalite, chalcedony, and petrified wood. These sources of beautiful, near-gem-quality stones were mined prehistorically and were the nearest source for the exotic cryptocrystalline flaked-stone artifacts found along the Sierra crest at the Morris Peak and Scodie Mountain archaeological sites (Garfinkel et al. 1980; McGuire and Garfinkel 1980).

In the northeastern portion of the El Paso Range is majestic Black Mountain, a large and dramatic, brown-black, Pleistocene basalt flow. Other volcanic rocks are found in the zone of contact with the Sierra Nevada in the adjacent desert areas and within the more northerly lava flows and cinder cones of the Coso Range. Red scoria, vesicular basalt, and pumice served as raw materials for groundstone implements including bowls, metates, and manos. Artifacts manufactured from these materials were recovered from sites along or in the near vicinity of the Sierra crest. The Cosos also contain a milelong (1.7-kilometer) seam of volcanic glass that was the principal source of obsidian, a favored material for the manufacture of flaked-stone tools.

Summary

The study area is situated near the interface of the Great Basin and the Mojave Desert. This is a land of extremes in relief, temperature, and moisture. Topography and climate dictate the range of flora and fauna that was exploited by the aboriginal populations.

All the archaeological deposits studied lie at elevations from 5,000 to 8,000 feet (1600–2500 meters). The 96 prehistoric site loci studied are located in or near a predominantly piñon woodland island, known as the Kern Plateau and Scodie Mountains located in the far southern Sierra Nevada. The presence of large stands of piñon provided a staple nut crop that was a principal food source and facilitated aboriginal occupation. The changing character of the piñon woodland environment over the last 5,000 years may have influenced the timing and intensity of subsistence-settlement patterns.

Chapter 3
Anthropological Background

Scope and Purpose

This chapter introduces anthropological methods for dating native languages and the societies they represent. Historical connections among languages and their geographic distributions suggest the direction and timing of the movements of their speakers in prehistory. Models accounting for these prehistoric movements, including the Numic expansion, are also presented. These models are evaluated and their implications for prehistory examined. Ethnographic overviews detail the character of aboriginal cultures and provide a window into the past, projecting historic subsistence-settlement patterns as well as sociopolitical and religious organization of native peoples back into the pre-contact era.

The chapter closes with a synopsis of the local archaeological record, outlining the prehistoric sequence. This brief cultural history reveals changing patterns of land use over time and the particular circumstances leading to the "ethnographic present." Such changes may be coincident with prehistoric population movements, including expansion, contraction, migration, and replacement.

Methods of Linguistic Prehistory

Comparative and other linguistic methods are used to reveal historical connections among languages. The resultant genetic classifications and loanword analyses may point to language homelands and can provide indications for the direction of population movements in the past. Glottochronology is a means of estimating rates of lexical change and the elapsed time since related languages diverged (Hymes 1960; Swadesh 1954). Since the technique relies on statistical methods, it is a form of lexicostatistics.

Lexicostatistics was originally developed by Morrris Swadesh and his colleagues in the 1950s (Gudschinsky 1956; Swadesh 1959). Its basic assumption is that certain common words are part of a basic vocabulary thought to be replaced at a slow, but fairly constant rate as language changes over time. The rate constant was originally derived from empirical determinations garnered through the study of replacement rates for written Indo-European languages (Diebold 1987).

The technique has been controversial and its basic assumptions often questioned. Yet despite these caveats, there seems to be little dispute that historical linguistics can demonstrate relationships among languages and provide a measure of the degrees of similarity (Foster 1996; Moratto 1984). Although glottochronology cannot supply unequivocal absolute dates for divergence of languages, it can provide a means for useful relative dating and that can be of great value to archaeologists. If prehistorians can assume a language-society-culture connection, they may be able to further refine the actual dating of prehistoric population movements through archaeological research.

The true role of historical linguistics and glottochronology is to provide the dendogram which can then be evaluated and tested independently with archaeological and DNA data. What I mean by this is that these techniques provide a means of developing the treelike model indicating the relative age and degree of connection between various languages. Reconstructed vocabulary is also used routinely as a basis for historical reconstruction and is referred to as the *Worter-und-Sachen* ('words and things') method. This reconstructed vocabulary can be used to develop a proto-vocabulary that points to a probable homeland for a language (Fowler 1972, 1983; Golla 2000).

Genetic Classification and the Distribution of Northern Uto-Aztecan Languages

The Uto-Aztecan language stock is distributed widely from Central America to southern Idaho (Fowler 1972). Most Great Basin languages belong to the northern branch of the Uto-Aztecan languages. The Northern Uto-Aztecan language group is composed of two familylike clusters, Numic and Takic and two single languages (linguistic isolates), Tubatulabal and Hopi (Foster 1996).

Genetic relationships and the spatial distribution characteristics for Numic languages were first identified by Kroeber (1907, 1925) building on earlier studies by Brinton (1891), Merriam (1904) and Powell (1891). Kroeber's earliest work identified Numic languages under the name of "Plateau Shoshonean" and divided them into three dialect groups rather than separate languages. Those groups included western Mono-Paviotso (Western Numic in current terminology, cf. Miller 1966), central Shoshoni-Comanche (Central Numic), and southern Ute-Chemehuevi (Southern Numic). The Kern River branch was recognized as having a single language group represented by Tubatulabal.

Lamb (1958), developing Kroeber's scheme further, proposed that the Numic language family contained two separate but related languages in each of its three branches. These were Mono and Paviotso for Western Numic, Panamint and Shoshoni-Comanche for Central Numic, and Kawaiisu and Ute (including Chemehuevi) for Southern Numic. He also identified Numic as most closely related to Tubatulabal. More recent works largely support the earlier classification schemes (Goss 1965, 1968; Kroeber 1959; Miller 1970, 1972, 1984, 1986; Miller et al. 1971; Steward 1937), and the linguistic diversity thesis initially provided by Lamb, has been largely upheld.

Hence, three languages (Mono, Panamint, and Kawaiisu) representing each of the three branches, are located in a small area of eastern California at the southwestern corner of the Great Basin. The other related languages extend in a great triangle with its apex in the southern Sierra Nevada and its base along the Rocky Mountain chain (Figure 1.4). This vast area includes the interior Great Basin (Nevada and Utah), Snake River Plain (in Idaho) and part of the Colorado Plateau. Significantly, there is little difference in dialects noted for the northernmost languages, while diversity is more marked in the southernmost area (Miller et al. 1971; Zigmond 1938).

Time Depths for Language Development and Separations

Estimates of language divergence based on lexicostatistical studies support the accuracy of the Lamb-Kroeber classifications (cf. Hale 1958; Swadesh 1954). This classification scheme provides a general sequence for the degree of relatedness of the various individual languages and the groups of languages as unified elements. The distribution of Numic languages (Figure 1.4) has long suggested their possible area of origin (Lamb 1958). Assuming that language change proceeds in a regular fashion through time, the greatest divisions in a language family are likely to reflect their greatest antiquity. The geographic location of these divisions would then indicate the area where the ancestral languages began to diversify (Sapir 1916, 1949; Swadesh 1954).

The center-of-gravity principle might locate the Numic homeland somewhere in the southwestern corner of the Great Basin (Lamb 1958; Miller 1966, 1986:102–103). Reconstructed plant and animal terms for a proto-Numic language also indicates a homeland of diverse elevation, in or near desert zones with access to substantial riparian resources, possibly at the interface of the southern Sierra Nevada and western Great Basin or in the Owens Valley (Fowler 1972, 1983).

The Tubatulabal language, being geographically and linguistically intermediate between the Numic and Takic languages, has probably been developing in place since its initial divergence from its Numic neighbors (Miller 1984). Lamb (1958) and Hale (1958), using the techniques and data from Swadesh (1954) and their own estimates, calculated a Numic/Tubatulabal separation on the order of 3000 to 2500 years ago. There is considerable debate concerning the precise age when proto-Numic separated into its three divisions, but about 1000 years later, Numic might have divided into its various subgroups. Linguists believe that approximately 1000 years ago or less the Numic languages spread rapidly across the Great Basin creating the historic distribution of languages recognized by anthropologists (Fowler 1972, 1983; Miller 1986).

Models and Mechanisms: Continuity Versus Replacement

Models of *in situ* Development: Contrary to the beliefs of many linguists and archaeologists, Aikens and Witherspoon (1986) and Aikens (1994) have suggested that Numic occupation of the Great Basin is long standing. They argue that the Great Basin has been continuously occupied by Numic peoples and their direct ancestors for the last 5000 years. They also argue that non-Numic groups occupied *only* the western and eastern Great Basin when the regional climate was relatively wet. During those periods, wetlands-adapted groups outcompeted Numic populations for favored areas. During times of increased dryness, those areas were abandoned and Numic populations expanded to fill the void.

Studies in the central Great Basin, particularly those in the Reese River Valley in central Nevada, support this suggested continuity in cultural pattern since about 2500 B.C. (Thomas 1971, 1972). Absorption or out-migration of non-Numic groups during dry cycles is suggested as characteristic of the edges of the Great Basin. Such a pattern of late prehistoric replacements is suggested to have occurred for the prehistoric cultural traditions known as the Anasazi, Fremont, and Lovelock. Western Numic and Southern Numic are suggested as coming into existence as Central Numic expanded in both directions. Such movements were characterized as a recurring pattern of "expansions and contractions that began as people first entered the Great Basin …" (Aikens and Witherspoon 1986:15). This model assumes that variations in hunter-gatherer adaptations are a function of environmental change rather than culturally distinct sequent populations pursuing varying subsistence-settlement strategies.

The Coso Style of rock art has figured rather prominently for researchers favoring a long-standing continuity of Numic languages in the southwestern Great Basin (cf. Grant et al. 1968; Pearson 2002; Whitley 1998). Whitley, building on the identification of historic horse-and-rider rock art depicted in some Coso Style paintings and rock drawings (Garfinkel 1982; Pearson 2002:80–83; Ritter et al.1982), sees an unbroken ethnic continuum in the region. He suggests that most rock drawings (petroglyphs) in the Cosos date only to the last 1000 years (within the time span and association of the historic Numic languages).

The association of Coso petroglyphs with Shoshonean peoples was first proposed by Grant et al. (1968). Grant and his colleagues originally supported that position because of the close correlation of the distribution of bighorn sheep and historic Numic languages. Since Coso Range rock art is, in fact, located in the general area identified by many as the proto-Numic homeland, some researchers believe that Numic peoples and their ancestors have lived there continuously for at least several thousand years (Pearson 2002; Warren 1984: 384; Whitley 1998:53–60).

Replacement Models: In contrast, Bettinger and Baumhoff (1982) draw upon optimal foraging theory to develop a model based on variations in adaptive strategies proposed for different hunter-gatherer cultural groups. Traveler strategies are linked with pre-Numic populations and processor strategies with the Numic. Travelers rely more heavily on high-quality resources of restricted distribution, such as bighorn sheep, necessitating greater mobility. Processors, in contrast, favor resources of lower quality, wider distribution, and higher processing costs, such as certain seed-producing plants and small game. The Bettinger-Baumhoff model ties most Great Basin rock art, except for historic pictographs and the "scratched" style, with pre-Numic populations. Bettinger also links with Numic processors a series of settlement-subsistence changes recognized in the archaeological record of eastern California. These changes include a shift to greater residential stability and a shift to permanent and semipermanent village life as well as increasing resource intensification with respect to greater use of seed and nut crops and small game.

Sutton (1986, 1987, 1994) alternatively proposes that the Numic spread was based largely on the control of critical resource patches. He hypothesized that Numic groups may have invaded areas largely devoid of settlements and taken control of certain concentrations of crucial resources, denying use to their non-Numic neighbors. These practices would imply defense of those areas either by their occupation or through overexploitation, leaving the areas depleted and virtually useless for others.

If the Numic people had larger social aggregates and more intense subsistence strategies, they would have been forced to move their settlements more often and leave the non-Numic populations continually disrupted. Sutton argues further that a general drying trend ca. A.D. 1000 may have been the initial trigger for Numic population movements out into the Great Basin from their homeland in eastern California. He suggests that warfare was a principal means by which Numic groups moved into and controlled resource areas. Sutton sees raiding and consistent outward expansion as characteristic of Numic culture and suggests that just such a pattern was at the heart of the ongoing Numic expansion.

Ethnography and Ethnogeography

Introduction

This ethnograpic overview provides a context for subsequent discussions tying the archaeology and prehistory of the study region to the historic linguistic groups in the area. This discussion focuses on cultural elements most relevant to the present study, including ethnogeography, territoriality, social and exchange relations, subsistence-settlement structure, sociopolitical organization, and religious concepts.

Ethnogeography deals with the way a culture perceives and defines its landscape. It considers what geographical areas comprised a homeland, what parts of the environment were selected for residence, which areas were used for hunting and gathering of principal foods, and what areas (if any) were defended against invaders or trespassers. Using the native significance of certain geographic areas and identifying the degree of territoriality provide clues to areas where boundaries may be sharp and others where they may be more diffuse.

Social and exchange relations provide parallel details regarding the level of interaction with neighbors and possible amity/enmity relationships. Subsistence-settlement structure, as recorded ethnographically, is key to understanding the character of land-use patterns and may lead to better predictions of the archaeological consequences of such activities. Sociopolitical organization helps in the understanding of the complexity of the native cultures, how populations were organized into various groups and how these groups might have been directed. Knowledge of religious beliefs, rituals, and ceremonial organization may be helpful for understanding the character of rock-art sites. Many researchers suggest that such sites are associated with shamanic rituals (Pearson 2002; Whitley 1998). Differences in the belief systems of the aboriginal peoples may be manifest in the subject matter and styles of their rock art.

Ethnogeography, Territoriality, and Tribal Relations

The study area lies at the junction of three different ethnolinguistic groups: the Tubatulabal, Kawaiisu, and Panamint Shoshone (Figure 1.11). Ethnographic data on the territories and boundaries of the three groups are presented in the primary references pertaining to these groups (Kroeber 1925; Steward 1937, 1938; Voegelin 1938; Zigmond 1938, 1981). These references generally agree with respect to the core territories for these groups but not with regard to the nuances of peripheral or boundary areas; there opinions vary.

Steward (1938:7, Figure 1, Page ix) identifies Tubatulabal territory a bit differently than Voegelin, Kroeber, and others in that he includes a portion of the western Mojave Desert just south of Little Lake within the territory. His map shows that the Tubatulabal may have routinely crossed the Sierra Nevada and entered the Mojave Desert for certain resources. Voegelin (1938) also documents such activities, including trips to the Indian Wells Valley to hunt antelope at the former community of Brown and to harvest blazing star (*Mentzelia* sp.) and chia (*Salvia columbariae*) seeds as well as Mariposa lily (*Calachortus kennedyi*) bulbs. Steward (1933) alludes to the possibility (cf. Ericson 1977:235) that the Tubatulabal may have once owned or controlled the Coso obsidian quarry, which, in Steward's estimation, would have been within their territory. Ethnohistoric sources indicate that the Tubatulabal traded and intermarried with the Panamint Shoshone, procured obsidian from them on expeditions into their territory, and collaborated in yearly communal deer hunts in the southern Sierra (Slater 2000:30).

It has been popular in anthropological circles to conceive of hunter-gatherers as peaceable and nonviolent with weakly held territories that were seldom defended and with little or no ownership of resources. Yet, evidence runs counter to this view, showing that territories were recognized, sometimes defended, and that resources were often exclusively held (Bettinger 1982; Irwin 1980; Kroeber 1925; Senett-Graham 1989). Ethnographic evidence hints that the Tubatulabal were not on good terms with the Panamint Shoshone and Kawaiisu and that some ongoing conflicts occurred. Steward notes that the Panamint Shoshone from Little Lake called the Tubatulabal *Nawavitc* or *Wavitx*, translated as "tough" or "mean" (Steward 1938:71–72). Voegelin (1938:49) also indicates that the Tubatulabal were engaged in hostilities to a greater extent than their Numic neighbors (Panamint Shoshone, Owens Valley Paiute, Kawaiisu).

One native consultant suggested that the Tubatulabal often fought with the Kawaiisu and the Panamint Shoshone. That consultant also stated that the Tubatulabal had waged a large battle with the Panamint Shoshone at Walker Pass and that another battle was fought with the Kawaiisu near their common border at Nichols' Peak. Several native consultants recounted details of another major battle at Haiwee Springs, where the Panamint Shoshone fought to defend their territory and killed many Tubatulabal (Irwin 1980:38–40.). Steward also notes a battle with an invading group at Coso Hot Springs where all the intruders were killed (Steward 1938:83). Smith (1978) indicates that the Tubatulabal engaged in warfare with all their neighbors and that their motivation for such conflicts was always revenge for prior hostilities. The Tubatulabal would take prisoners and scalps and kill men, women, and children during battles that lasted one or two days.

The Numic groups in the study area, the Kawaiisu and Panamint Shoshone, were far more amicable with one another than the Tubatulabal were with the former groups, although the Tubatulabal seemed to allow incursions onto lands they rarely used. During the early 1860s both the Kawaiisu and Panamint Shoshone established settlements on the eastern end of the South Fork Valley at the mouth of Spanish Needle Creek, just west of Walker Pass, in an effort to escape conflicts with Euroamericans in their own territories (Voegelin 1938:51). Voegelin's hamlet map (1938: Figure 11) of ethnographic villages ca. 1860 identifies Village Sites 1 and 2 as Panamint Shoshone and 3 as Panamint Shoshone-Kawaiisu.

Grosscup's examination of C. Hart Merriam's notes (1977) indicates that the Tubatulabal's eastern border was probably the crest of the Sierra near Canebrake Creek with Walker Pass mutually occupied by both the Panamint Shoshone and Kawaiisu. Zigmond (1938) generally agrees but suggests that the Kawaiisu alone controlled Walker Pass. Zigmond's ethnographic data on the Kawaiisu focus on their settlements and activities which were mainly in a "core area" in the Tehachapi Mountains where winter settlements were located. Yet multiple sources (Driver 1937; Irwin 1980; Senett-Graham 1989; Steward 1937, 1938:93, Figure 7) say that the Kawaiisu groups were strongly allied with the Panamint Shoshone and had distinct districts or subgroups occupying exclusively desert territories (cf. Underwood 2004).

Districts: When considering the territoriality of the Panamint Shoshone and the Kawaiisu, it is increasingly evident that their societies were organized into geographical units or districts. These districts were relatively exclusive, largely nonoverlapping geographical areas associated with key water sources and major village settlements. District organization was loose enough to allow for residence change, and intermarriage between districts was necessitated. Yet Native American consultants verified that they did not randomly venture into other districts (Irwin 1980:xiii). Men and women seldom traveled into districts that were not within their family's home range except for special festive occasions or group hunts (Irwin 1980:xiii; Sennett-Graham 1989:25).

Steward identified seven districts among the Panamint Shoshone and Kawaiisu (Steward 1937, 1938). Those districts included (1) the area north of Lida, (2) Beatty and the Belted Range (in western Nevada), (3) Saline Valley, (4) Little Lake and the Coso Range, (5) Panamint Valley, (6) northern Death Valley, and (7) southern Death Valley. Driver (1937) identified five subgroups of the Panamint. Grosscup (1977), providing data from Merriam, reported six divisions for the Panamint Shoshone. The Panamint Valley and southern Death Valley districts were composed of almost equal numbers of Shoshone and Kawaiisu. The southern portion of Panamint Valley was predominantly Kawaiisu. When borderlands were occupied, it was, in fact, common that settlements would include people speaking related but different languages. Kawaiisu speakers were part of Steward's Koso (*Pawo'nda*) or Little Lake district that included the region of the Coso Range, Rose Valley, Little Lake, Olancha, Darwin, Walker Pass to Owens Lake, and part of the far southern Sierra Nevada (Steward 1937, 1938; see also Voegelin 1938).

Tubatulabal

Coverage, Name, Territory, Population Estimates, Village Locations: Ethnographic material on the Tubatulabal is found in Kroeber (1925), unpublished notes by John Peabody Harrington (1934) and C. Hart Merriam (1937–1938), and treatments by Voegelin (1938) and Smith (1978). Voegelin's monograph is the most substantive and exemplary, providing great detail in most matters. An important ethnohistoric source for the Tubatulabal is B. Powers (1971, 1974,1981). His books provide significant detail regarding the character of protohistoric and early historic Euroamerican-Indian social and economic interaction, chiefly in the South Fork valley of the Kern River. These Native people referred to themselves as Tubatulabal, which translates as "piñon pinenut eaters" (Merriam 1904).

Traditional Tubatulabal territory is centered in the far southern Sierra and includes the region naturally drained by the Kern River. That territory begins near Mount Whitney, and terminates below the confluence of the two forks in the Kern River Canyon, at a place just above the rapids at the end of the Lower Kern River Canyon northeast of Bakersfield (Smith 1978:437). Estimates place the Tubatulabal precontact population at between 500 and 1,000 (Kroeber 1925:608). The Tubatulabal were composed of three distinct bands: *Tolowim*, *Pahkanapil*, and *Palegewan* (Voegelin 1938). Each occupied geographically demarcated areas during the winter. The *Tolowim*, called *Bankalachi* by the Yokuts, were associated with the Hot Springs Valley and were closely allied with the Yokuts. The *Palegewan* were in the Kern River Valley, and the *Pahkanapil* inhabited the South Fork Valley of the Kern where most of the population was aggregated (Kroeber 1925; Smith 1978; Voegelin 1938).

Subsistence and Seasonal Round: The Tubatulabal had a relatively lush environment, with both central Californian and Great Basin resources (Voegelin 1938). They lived in an area encompassing riverine, piñon-juniper, and high Sierran environments, and their homeland was rich in resources. Importantly, they had access to two major dietary staples, piñon nuts and acorns. Fish were next in economic importance.

From February to May, the Tubatulabal obtained most of their food within the South Fork Valley. Stalks of our Lord's candle (*Yucca whipplei*), immature Joshua tree (*Yucca brevifolia*) fruit and various bulbs were gathered. Large and small game were taken, including geese that were usually available in March. The Tubatulabal fished near the confluence of the South Fork with the main channel of the Kern River. Piñon nuts and acorns, stored from the previous season's harvest, added to the diet (Voegelin 1938).

During May, the people gathered seeds on the lower foothills and valley floor. Grey pine (*Pinus sabiniana*) nuts and juniper (*Juniperus occidentalis*) berries were harvested. Large and small game animals, especially rabbit, continued to be hunted in the valley and foothills. Some fishing took place in the main channel of the Kern River. Occasionally families would venture across the Sierra into Indian Wells Valley for seeds, bulbs, and small game (Voegelin 1938).

By June stored foods were normally exhausted, but if the season was late, seeds might still be gathered in the Kern River Valley. At this time of year, they gathered tule roots around springs, and conducted rabbit drives. They also hunted large game and small birds. From July to mid-August they dug rush roots near streams, and gathered manzanita berries in the mountains. Fish were netted in the Kern. Mussels were gathered near Kernville, and rabbits were hunted on the Kern River Valley floor (Smith 1978; Voegelin 1938).

In mid-August to September, the piñon season began at the lower elevations and families would begin to harvest nuts for several days at a time (Butterbredt 1948; Voegelin 1938). Near the piñon grounds, they trapped small game and hunted deer. They also gathered juniper berries and fish were stupefied with a certain fish poison as streams became lower. From September to mid-October piñon nuts were gathered at higher elevations. Again, hunting of small game and deer took place. From mid-October to mid-November families moved back to their villages carrying burden baskets laden with piñon nuts to be stored for winter. If a local crop was especially good and the winter not too severe, families might have relocated their villages and stayed in the area of the nut harvest. In the late fall in the Greenhorn Mountains, acorns were harvested, and deer and rabbits were taken (Voegelin 1938). From mid-November to February, subsistence was based largely on stored foodstuffs. Supplementing the dried foods were fish speared from the river and game that was hunted or trapped (Voegelin 1938).

Of major importance to the present study was the gathering of piñon nuts and hunting of small and large game at the piñon grounds. The piñon harvest at higher elevations began in fall. Nuts were gathered and cached at this time. Two types of piñon settlements are indicated ethnographically: a large camp with a corral-like brush enclosure housing a number of extended families, and a more temporary camp having a shelter of poles and brush used by a single family during brief stays.

Piñon nuts may be harvested at the green- or brown-cone stage, and much has been made of this distinction (Bettinger 1976; Bettinger and Baumhoff 1982; McGuire and Garfinkel 1976). The ethnographic details seem to point to an exclusive green-cone harvesting method for the Tubatulabal. Piñon nuts seem to be most productively harvested when green, when the cones were full grown but their scales unopened (Bettinger and Baumhoff 1982). Men and boys would knock the cones from the trees with long poles. When enough cones had been gathered, they were taken to camp, dumped onto a bed of sage (*Artemisia tridentata*), and set on fire (Voegelin 1938). Heat from the fire caused the cones to dry out and their scales to open. The cones were then allowed to cool and the nuts shaken out, picked out by hand, or winnowed from the dirt. Nuts not taken back to the village were cached in circular pits covered by rocks, grass, and small stones. These caches were returned to in the winter as needed (Voegelin 1938).

Social and Political Organization: The Tubatulabal were organized into three semi-independent, politically differentiated bands, each with its own chief. Associated with each band were several hamlets or permanent winter villages, each included from two to six extended families. The Tubatulabal were territorially based and claimed property rights as a community recognizing certain geographical boundaries marking their territory. Each band's chief acted as a counselor, arbitrator, and representative of his band. His responsibilities included leadership in war and peace, negotiation of internal disputes, and admonishment or punishment of shamans suspected of malicious or injurious behavior (Voegelin 1938).

The band chief held an appointed office, and upon his death, an assembly was called and a new man was chosen. The next most important position in Tubatulabal society was the "clown/dance manager," a man who inherited this office from his father (Voegelin 1938). The position served to create levity at ceremonies and when appropriate was also instrumental in calling for a change in band leadership.

Religion, Cosmology, and Group Ceremonies: Religious life for the Tubatulabal centered around the concept of a dying benefactor and was related to their use of jimsonweed (*Datura* spp.). The Tubatulabal believed the world was inhabited by various supernatural spirits in both human and animal forms (Voegelin 1938). Such beings sometimes figured as a shaman's helper and were always treated with reverence and a bit of trepidation. Tubatulabal shamans included both men and women and served as doctors in curing ceremonies or, when malevolent, as witches. Other Tubatulabal shamans included weather shamans or rain doctors who were able to produce rain when needed. Bear shamans were uncommon, but occasionally shamans obtained bears as guardian animals. Vision seekers and shamans sometimes took rattlesnakes as guardians and those who had such associations could cure rattlesnake bites (Voegelin 1938).

Group ceremonies among the Tubatulabal included an annual mourning ceremony for the dead. When the ritual took place, images and possessions of the deceased were burned. A ceremony, identified as the "little fiesta" or face-washing ceremony, was conducted for the survivors of a deceased person before they were able to resume eating meat (Voegelin 1938). Young men and women, using jimsonweed to obtain visions and spiritual guardians, performed a special ritual.

Panamint Shoshone

Documentary Coverage: Ethnographic accounts of specialized aspects of the Panamint Shoshone include the early works by Nelson (1891), Coville (1892), and Dutcher (1893). More recent overviews and refinements have been published by Grosscup (1977), Kelly and Fowler (1986), and Thomas et al. (1986). Steward (1937, 1938) provides the most comprehensive picture of aboriginal life for these groups. Kroeber (1925) includes a brief summary treatment. Important works by Irwin (1980), Sennet-Graham (1989), and Slater (2000) provide significant additional information, including historic accounts of the changes in traditional native culture that occurred after prolonged influence from the dominant Euroamerican economy.

Name, Territory, Population Estimate, Village Locations: The appellations Coso (alternatively Koso) and Panamint Shoshone have been used to refer to the aboriginal people living in the Coso Range and the surrounding areas. The Little Lake or *Kuhwiji* district was located nearest the Kern Plateau and the crest of the Sierra. The *Kuhwiji* district was large and encompassed an area of almost 1000 square miles consisting of four loosely interrelated villages. The village at Little Lake was called *Pagunda* and had 50 or 60 residents in 1870. Coso Hot Springs or *Mua^ta* had a population of more than 100. Cold Spring was located about five miles (8 kilometers) south of Darwin. The Olancha village was situated along the northern boundary of the district and included Northern Paiute speakers. Voegelin (1938) mentions that three Panamint Shoshone villages were located near Walker Pass not far from Canebrake Creek but that these villages may have been recent historic intrusions. Grosscup, using the notes of C. Hart Merriam, suggests that the Panamint Shoshone and the Kawaiisu mutually occupied the Walker Pass area west of the crest of the Sierra near Canebrake Creek. Estimates of population density were between 0.06 and 0.03 persons per square mile (0.002–0.001 person per km^2), or 16 to 30 square miles per person (500–1000 km^2 per person) (Steward 1938).

Seasonal Round: The Panamint Shoshone environment was much less productive than that of their neighbors, the Tubatulabal, especially with reference to the availability of water. The Coso area is just one valley system away from Death Valley — one of the driest places on earth. Hence, the Shoshone were much more residentially mobile than the Tubatulabal. The Shoshone seasonal round varied depending on the relative abundance and location of key food sources.

The Panamint Shoshone subsisted by generalized foraging. During the winter, people moved to valley-floor villages next to streams, probably along the eastern scarp of the Sierra (Steward 1938). There they occupied pit houses and lived mainly on stored seeds and nuts and they hunted rabbits. In spring, some families moved to Haiwee Spring to gather greens. In late spring and early summer, some people moved to Cold Spring to hunt rabbits. At the same time, other families would travel and convene for communal antelope drives. Such drives were held at Brown (near the modern town of Inyokern), at the southern end of Owens Lake, or at the north end of Saline Valley (Steward 1938).

The drives near Brown might have involved cooperative efforts with neighboring groups including the Tubatulabal or the Saline Valley Shoshone. During the middle of summer, families would travel to Saline

Valley or sometimes Death Valley to gather mesquite beans. In late summer, people moved throughout the Coso Range to gather plant foods. In the fall most families traveled to the productive piñon grounds in the Cosos (and probably also the nearby piñon grounds of the southern Sierra). Alternatively, if the crops failed, families would move to the Panamints (Steward 1938). Some families would travel to Owens Lake to hunt waterfowl. Fall was also the time for the large communal rabbit drives. Important supplements to the largely vegetal diet were the hunting of bighorn in the Cosos, deer and bighorn in the Sierra, and fishing in the Owens River and Little Lake (Steward 1938).

Social Organization: As Steward often declared, the Great Basin Shoshoneans possessed one of the simplest cultures in North America and in some respects the entire world. A corollary to this simplicity was that they were characterized by the most basic level of sociopolitical organization, a band-level society. These localized bands were composed of groups of related and cooperating extended families that lived within a recognized territory or district, shared a sense of common membership and identification, and were directed by a local leader. These chiefs or headmen organized the annual rabbit drives and occasional communal piñon harvests, influenced settlement decisions, officiated at and organized the annual Round Dance (see next section), resolved interpersonal conflicts, punished thieves, and helped lead in decisions regarding intergroup conflicts (i.e., warfare).

Religion, Cosmology, and Group Ceremonies: The Panamint Shoshone had little in the way of embellishment in their material culture, no elaborate ceremonialism, and only slight ritual or religious activities (Steward 1968:ix). Guardian-spirit beliefs, crisis rites, mythology, and shamanism formed the core of religious concepts (Hultkrantz 1986). Shamans acted as religious functionaries and healers. They served as doctors to cure sickness, which was thought in some cases to come from malevolent individuals (shamans and non-shamans), or the violation of certain taboos. Communal religious activities were few but occasionally, when group rituals were practiced, they included the Fall Round Dance, also known as the Circle Dance. The Round Dance included a first fruit rite following the annual piñon harvests and communal rabbit drives. This Fall festival appears to have had a world renewal component and an association with the promotion of the growth of plants (Hultkrantz 1986: 634; Steward 1941:267).

Kawaiisu

Documentary Coverage: Information on the Kawaiisu is scattered in many fragmented accounts (cf. Underwood 2004). The earliest facts appear in the scant details provided by S. Powers (1877). Local historians have added important ethnohistoric and ethnographic details (Barras 1973, 1976, 1984; B. Powers 1981; Walker 1971). Historians have documented the troubled period of missionization, forced relocation, and intense conflict with the government and settlers (Boyd 1972; Chalfant 1933). Precontact lifeways are also treated in Cappannari (1950, 1960), Kroeber (1925), and Steward (1938). Yet the most detailed and thorough studies are the various works by Zigmond (1938, 1941, 1971, 1972, 1977, 1978, 1980, 1981, 1986; Zigmond et al. 1991). Unfortunately, Zigmond's works lack some important details necessary for the archaeological reconstruction of precontact life, including the location and character of principal villages and a thorough discussion of Kawaiisu subsistence-settlement patterns. Consequently, prehistorians are forced to patch together a coherent picture as best as possible given these limitations.

Territory and Village Location: Zigmond (1986) identified the Kawaiisu as centered in the far southern Sierra Nevada, principally in the Piute and Tehachapi Mountains. Yet Steward (1938) also assigns a number of Kawaiisu groups to the southern portions of Panamint Valley and Death Valley. Steward notes that south of Ballarat, Panamint Valley was largely inhabited by Kawaiisu speakers, and that their principal village, called *Ha:uta* (Village 42 in Steward's Figure 7), was at Warm Springs. One of Zigmond's native consultants confirmed that the Kawaiisu would travel across Indian Wells Valley and into the Argus Range. Grosscup (1977), using the notes of C. Hart Merriam, attests that the Kawaiisu claimed the territory near Walker Pass, and Voegelin further attributes a village in that area (at least during early historic times, ca. 1860) to the Kawaiisu. That village was situated at an unnamed spring near the mouth of Spanish Needle Creek.

Subsistence and Seasonal Round: Zigmond (1986) mentions that Kawaiisu territory was not richly endowed with subsistence resources. He also states that at times the Kawaiisu verged on starvation or suffered from lack of provisions. Steward (1938:84) recounts that the Kawaiisu from Panamint Valley would harvest mesquite beans (*Prosopis juliflora*) at their Warm Springs winter village. They would also travel to higher elevations to gather seeds and piñon nuts and hunt bighorn sheep. Families would also travel to the Argus Range and Coso Mountains for chia and bunch grass. Groups living in the Tehachapi Mountains had both acorns and piñon nuts to gather during the fall. In the spring, various seed-producing plants were gathered. Among the most important were rice grass (*Achnatherum hymenoides*), tick seed (*Coreopsis* spp.), blazing star (*Mentzelia* spp.), and chia (*Salvia* spp.). Some fishing was done but few good fishing steams were available. Rabbits were hunted communally.

Social Organization: Social organization centered on the family group with little supra-familial political organization. Chiefs were known, but no single individual united the Kawaiisu as a whole. Leaders were simply individuals who possessed personal wealth but no real coercive authority. Yet they supervised feasts and bore much of the expense for such group ceremonies (Zigmond 1986).

Religion, Cosmology, and Group Ceremonies: Group ceremonies among the Kawaiisu included an annual mourning ceremony where images and possessions of the deceased were burned and a ritual when boys and girls a few years after puberty used jimson weed to obtain visions and spiritual guardians. The religious world of the Kawaiisu was similar to that of their neighbors, the Panamint Shoshone, in having guardian-spirit beliefs, elaborate mythology, and shamanism. Three kinds of shamans were known: the curing shaman diagnosed and healed illness, evil shamans might attack their victims through supernatural agents and cause them to become ill or die, and weather shamans were a specialty of the Kawaiisu and could produce rain or snow (Voegelin 1938; Zigmond 1977, 1980, 1986).

Archaeological Background

Archaeological investigations in the far southern Sierra Nevada have been rather sporadic. Most of the research was done in the 1980s when the Pacific Crest Trail was developed. Prior research has been synthesized by McGuire and Garfinkel (1980), Schiffman and Garfinkel (1981a), and Moratto (1984). Prehistoric use of the region may have begun as early as 13,000–13,500 years ago, based on the identification of a few isolated finds of fluted Clovis points (Figure 3.1a) (Dillon 2002; Glennan 1971; Zimmerman et al. 1989). Nevertheless, little in the way of archaeological material has dates earlier than 3000 B.C., and most material falls within later prehistoric periods. With prior studies as a basis, an outline of the area's cultural history can be developed following the chronological periods specified in the local sequence.

Kennedy Phase (6500–11500 B.C.): The first known use of the Kern Plateau is represented by the Kennedy Phase. Large lanceolate concave base points are characteristic of this period (Figure 3.1b). These projectile points compare favorably with Great Basin Concave Base series found in the desert areas to the east. Great Basin Concave Base points have been found in contexts associated with radiocarbon

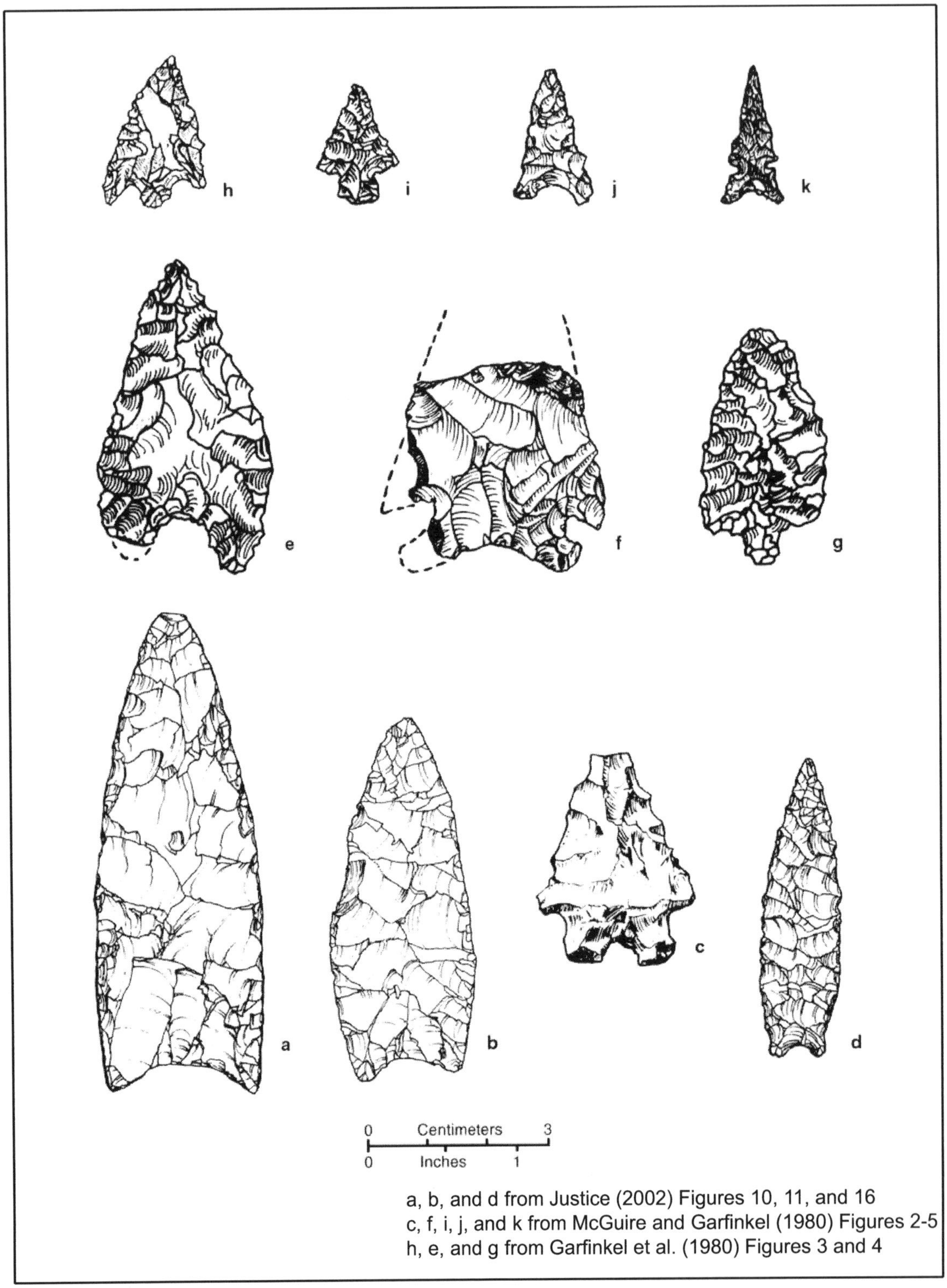

Figure 3.1 Time-Sensitive Projectile Points.

a. Clovis, b. Great Basin Concave Base, c. Pinto/Little Lake, d. Humboldt Concave Base, e. Humboldt Basal-notched, f. Elko Corner-notched, g. Gypsum, h. Eastgate, i. Rose Spring, j. Cottonwood, k. Desert Side-notched. All actual size.

dates and obsidian hydration measurements placing them in the late Pleistocene and early Holocene eras. With the exception of these early-style points (n = 2) and a small number (n = 7) of large hydration measurements on Coso obsidian (>10.1 microns), little in the way of archaeological materials have been recovered with which to reconstruct the prehistoric lifeways dating to this time.

Lamont Phase (6500–1200 B.C.): Various split-stem points of the Pinto/Little Lake series characterize this period (Figure 3.1c). These dart points, which are relatively rare (n = 9) in the study area, are thought to represent brief occupations and occasional use of the upland areas by big-game hunting parties, probably originating from base camps on the western fringe of the Great Basin desert. Occasional exploitation of various plant resources probably took place during this period as well (McGuire and Garfinkel 1980).

Canebrake Phase (1200 B.C.–A.D. 600): The hallmarks of this period are Humboldt, Elko, and Gypsum series projectile points (Figure 3.1d, e, f, and g). Hunting continued as well as new activities associated with substantial trans-Sierran trade. Locations of obsidian-reduction sites or lithic workshops suggest where the production of obsidian bifaces occurred. These bifaces were most likely made for exchange with groups to the west (McGuire and Garfinkel 1980; Schiffman and Garfinkel 1981a).

Continuing sporadic use of economic plant resources also characterizes this period. Piñon nut exploitation may have begun by 500 B.C. to A.D. 1, perhaps as a result of the ameliorating climatic conditions of the Medithermal Period and the emergence of larger, more productive stands of piñon pines (McGuire and Garfinkel 1980).

Sawtooth Phase (A.D. 600–1300): During this period, a transition is recognized from the use of the atlatl and dart to the bow and arrow indicated by Eastgate and Rose Spring points (Figure 3.1h and i). Aboriginal use of the area dramatically increased in both the size and number of archaeological sites and in the presumed intensity of occupation. Exploitation of piñon pine intensified as demonstrated by many small rock-ring features that served either as the bases for temporary brush structures at piñon camps or, more often, as rock-lined storage facilities for piñon caches. Individual hunting camps declined in use, and lithic workshops were no longer visited (McGuire and Garfinkel 1980).

Chimney Phase (A.D. 1300–Historic): This final cultural period represents ethnographic occupation by the Tubatulabal, Kawaiisu, and Panamint Shoshone. The greatest numbers of sites and artifacts found on the Kern Plateau date to this period. Desert Side-notched and Cottonwood arrow points are characteristic (Figure 3.1j and k). Brownware ceramics, imported soapstone beads, and pictographs date to this time frame, as do many sites associated with systematic and intensive piñon exploitation (McGuire and Garfinkel 1980).

Summary

Northern Uto-Aztecan languages include Tubatulabal, a linguistic isolate, and two Numic languages: Kawaiisu (Southern Numic) and Panamint Shoshone (Central Numic). Based on the distribution of Numic languages, the Numic homeland is thought by some to be in the vicinity of the far southern Sierra and western Mojave Desert or in the Owens Valley. Time depth estimates for these languages vary. Based on lexicostatistical measures, Tubatulabal is the oldest at 2000 to 3000 years, and the Numic neighbors may have differentiated about a millennium later. The expansion of Numic into the Great Basin may have occurred ca. A.D. 1000.

Theorists have advanced contrasting models regarding the ethnic identification and population movements of the prehistoric occupants of the study area. Aikens and Witherspoon suggested that Numic use of the Great Basin is of long standing, that Numic groups occupied the area for the last 5000 years, and that they displaced pre-Numic groups when climatic conditions became more arid. Bettinger and Baumhoff differ and believe that population replacement was driven by competing adaptive strategies. Sutton favors population replacement but sees warfare as a principal factor in the process. Others, considering the archaeological record (including the dating and styles of Great Basin rock art) argue for longterm Numic continuity.

It can be argued that the Tubatulabal can best be seen as typically Californian in cultural orientation. These people had substantial population numbers and formalized social structure, occupied semipermanent villages, and inhabited a richly endowed natural environment with abundant foodstuffs. Their subsistence-settlement activities were concentrated in the valleys and highlands of the far southern Sierra. In contrast, the Kawaiisu and Panamint Shoshone are more typical of certain Great Basin peoples having highly mobile family bands. Although the Kawaiisu held tenaciously to the relatively wellwatered slopes of the Tehachapi Mountains, the region they inhabited had few streams, and a large portion of their territory incorporated areas of desert. The Panamint Shoshone territory lay almost entirely within the drier regions of the Great Basin, east of the Sierra. The Panamint and Kawaiisu people had little in the way of artistic embellishments in their material culture and only slight religious ritual. The Tubatulabal had somewhat more complex religious systems with a variety of shamans and several forms of group ceremonies.

The sequence of prehistoric occupation as presently known begins about 6000 B.C. and continues through the historic era. Earlier human activity occurring prior to this is poorly visible. The archaeological record suggests that aboriginal activities in the area were generally focused on hunting forays, seasonal piñon exploitation, trans-Sierran travel, trade of obsidian, and, in good piñon years, more lengthy residential occupations. Significant changes occurred over the course of the 8000 years of documented prehistory. Prehistorians have identified varied occupations of the region, including changes in technology, the patterns of large-game hunting, trade in obsidian, and the development of systematic and intensive piñon exploitation.

Chapter 4
Chronology

Scope and Purpose

Accurate dating of prehistoric components is a prerequisite for testing hypotheses about prehistoric population movements or the ethnic attribution of archaeological assemblages. Knowing when sites were occupied provides a structure for inquiry regarding cultural variability or changes in land use over time. Knowing when events occurred can easily change our understanding of the nature of the events and their interpretation. Chronology is thus the foundation upon which explanations of prehistory must be built.

Archaeological investigations on and near the Kern Plateau allow for temporal ordering of many occupational components (Table 4.1). Dating is based on radiocarbon assays, source-specific obsidian hydration measurements, and time-diagnostic artifacts (projectile points, shell, stone and glass beads, and pottery). This chapter presents a chronological overview, provides new perspectives on the dating of prehistoric sites in the study area, presents the rationale for these interpretations, and addresses local occupational trends over time.

Introduction

Obsidian hydration dating is the primary means for placing the archaeological components within a time sequence. A summary of the technique is provided in the section on Obsidian Hydration Dating later in this chapter. The current sample of obsidian hydration measurements includes 475 hydration rim readings on projectile points and debitage from the 69 investigated sites (Table 4.1). All of these obsidian specimens were further analyzed to determine their geological source.

In almost every instance, X-ray fluorescence (XRF) analysis identified these obsidian artifacts as volcanic glass from the Coso Volcanic Field. To date, only five obsidian artifacts from the study area have been determined to come from other known or unknown volcanic glass sources. Over 99 percent of all the obsidian identified originated at the Coso sources. All of the source determinations were made in the 1980s, and at that time the Coso Volcanic Field was recognized as a single discrete chemical source.

Since that time, studies have identified four main subsources: Sugarloaf Mountain, West Sugarloaf Mountain, Joshua Ridge and West Cactus Peak (Ericson 1977; Ericson and Glascock 2004). Other subsources may also be present. The subsources represent different chemically distinct flows. Recently both Gilreath and Hildebrandt (1997) and Eerkens and Rosenthal (2004) have suggested that all four subsources produce hydration rims at a similar rate. These authors also found no statistically significant differences when they compared hydration measurements from various subsources. These findings allow us to date artifacts of Coso obsidian regardless of the particular Coso subsource represented.

Chronology

A chronological scheme for the Kern Plateau was originally developed by Garfinkel et al. (1981). That scheme has generally been retained, with some modifications, in the present work. The local cultural

Table 4.1 Distribution of Temporal Indicators by Site.

Site	Hydration Measure-ments	Classifiable Points	Radiocarbon Assays	Potsherds	Glass Beads	Shell Beads	Stone Beads	Other
KER-1269/SP4	7	2	0	0	0	0	0	0
KER-1270/SP5	2	0	0	0	0	0	0	0
KER-1971/SP6	2	0	0	0	0	0	0	0
KER-1296/SP8A	18	1	0	0	0	0	0	0
KER-1297/SP8B	6	3	0	75	1	0	0	0
KER-1298/SP8C	29	12	2	29	234	0	0	0
KER-1299/SP8D	7	2	0	1	0	1	0	0
KER-1299/SP8E	2	0	0	0	0	0	0	0
KER-1299/SP8F	8	1	2	1	0	0	0	0
KER1272/SP9	0	0	0	0	0	0	0	0
KER1273/SP10	1	0	0	0	0	0	0	0
KER1274/SP11	0	0	0	0	0	0	0	0
KER-1275/SP13	1	1	0	62	0	0	0	0
KER-1276/SP14A	1	0	0	0	0	0	0	0
KER-1276/SP14B	1	1	1	0	0	0	0	0
KER-1277/SP15	8	1	2	0	0	0	0	0
KER-1278/SP16	2	1	0	0	0	0	0	0
KER-1279/SP18	10	0	0	0	0	0	0	0
KER-1280/SP19	1	0	0	0	0	0	0	0
KER-1281/FS56	11	5	0	0	0	0	0	0
KER-1282/FS64	2	0	0	0	0	0	0	0
KER-1283/FS65	9	5	2	2	1	0	0	0
KER-1284/FS66	3	1	0	0	0	0	0	0
KER-1285/FS75	4	0	0	0	0	0	0	0
KER-1286/FS76	12	2	3	41	3	1	0	1*
KER-748/PCT1	2	3	0	0	0	1	0	0
KER-744/PCT2	0	0	0	1	0	1	0	0
KER-743/PCT3	1	3	1	0	0	0	1	0
KER-742/PCT4	0	0	0	1	0	1	0	0
KER-741/PCT5	0	0	0	1	0	0	0	0
KER-747/PCT7	0	0	0	0	0	0	0	0
KER-746/PCT9	0	0	0	0	0	0	0	0
KER-738/PCT10	0	0	0	0	0	0	0	0
KER-745/PCT11	0	0	0	0	0	0	0	0
KER-737/PCT12	0	2	0	0	0	0	0	0
TUL-484/PCT13	0	4	0	0	0	0	0	0
TUL-483/PCT14	0	7	0	0	5	0	1	0
TUL-482/PCT15	3	4	0	0	3	3	0	0
TUL-481/PCT16	3	9	0	0	13	10	0	0
TUL-480/PCT17	0	0	0	0	0	0	0	0

*Phoenix button

Table 4.1 Distribution of Temporal Indicators by Site (continued).

Site	Hydration Measure-ments	Classifiable Points	Radiocarbon Assays	Potsherds	Glass Beads	Shell Beads	Stone Beads	Other
TUL-487/PCT18	0	0	0	0	0	0	0	0
TUL-485/PCT 19	0	1	0	0	0	0	1	0
TUL-488/PCT20N	23	26	0	3	0	7	13	0
TUL-488/PCT20S	0	0	0	0	0	0	0	0
TUL-489/PCT21	0	0	0	0	0	0	0	0
TUL-629/KR39	23	5	0	1	18	10	6	0
TUL-621/KR41	33	13	0	4	6	0	7	0
TUL-620/KR42	0	4	0	0	0	0	7	0
TUL-619/KR43	7	2	0	0	0	0	0	0
TUL-618/KR44	0	0	0	0	0	0	0	0
TUL-630/KR46	7	0	0	0	0	0	0	0
TUL-617/KR48	0	1	0	0	1	0	0	0
TUL-628/KR49	0	0	0	0	0	0	0	0
TUL-616/KR50	1	1	0	0	0	0	0	0
TUL-767/KR53	0	1	0	0	0	0	0	0
TUL-636/KR57	7	4	0	0	0	0	0	0
TUL625/KR60	3	1	0	0	3	0	0	0
TUL-623/KR64	3	0	0	0	0	0	0	0
TUL-634/KR71	2	1	0	0	0	0	0	0
TUL-632/KR73	0	1	0	0	1	0	0	0
TUL-877/RB1	13	4	3	0	0	0	0	0
TUL-878/RB2	0	0	0	0	0	0	0	0
TUL-879/RB3	27	32	8	19	0	2	4	0
TUL-880/RB4	0	0	0	0	0	0	0	0
TUL-881/RB5	0	0	0	0	0	0	0	0
TUL-882/RB6	0	3	0	0	0	0	0	0
TUL-883/RB7	0	0	0	0	0	0	0	0
TUL-884/RB8	5	1	0	0	0	0	0	0
TUL-885/RB9	0	0	0	0	0	0	0	0
TUL-886/RB10	0	0	0	0	0	0	0	0
TUL-887/RB11	4	2	0	0	0	0	0	0
TUL-890/RB12A	5	1	1	60	0	0	0	0
TUL-890/RB12B	7	2	1	1	0	0	0	0
TUL-890/RB12C	3	0	0	0	0	0	0	0
TUL-891/RB13	17	0	1	11	0	0	0	0
TUL-888/RB17	0	0	0	0	0	0	0	0
TUL-889/RB18A	10	7	0	0	0	0	1	0
TUL-889/RB18B	1	0	0	0	0	0	0	0
TUL-889/RB18C	5	3	0	0	0	0	0	0
TUL-889/RB18D	2	2	0	0	0	0	0	0
TUL-894/RB19	3	2	0	0	0	0	0	0

Table 4.1 Distribution of Temporal Indicators by Site (continued).

Site	Hydration Measure-ments	Classifiable Points	Radiocarbon Assays	Potsherds	Glass Beads	Shell Beads	Stone Beads	Other
TUL-895/RB20	1	1	0	0	0	0	0	0
TUL-511/05-13-36	2	3	0	70	0	0	2	0
TUL-896/KM1	14	7	0	0	0	0	0	0
TUL-897/KM2	8	2	0	0	0	0	0	0
TUL-898/KM3	46	2	1	1	0	0	3	0
TUL-899/KM4	37	8	0	0	0	0	0	0
TUL-909/KM14	0	1	0	0	0	0	1	0
Totals	475	222	28	396	282	24	46	1

sequence was derived from the generally accepted temporal divisions for Great Basin projectile point types and the time periods developed by Bettinger and Taylor (1974) and Warren (1984). Period names are consistent with prior treatments (Garfinkel et al. 1980; McGuire and Garfinkel 1980) with the exception of the addition of a new Kennedy Period (Table 4.2). The hinge points for these periods have been adjusted based on current age estimates of the point forms.

The chronological evaluation of the archaeological material takes into account depositional contexts, the vertical and horizontal location of artifacts, the relative homogeneity of cultural assemblages from a particular area, and other relevant information. Discrete features and spatially concentrated scatters of surface artifacts within a site were evaluated for the possibility that they represented a single period occupation. Obsidian hydration measurements and radiocarbon assays help determine whether a given deposit represents a discrete period of cultural activity.

This chapter presents hydration rim measurements by site and component. Outlying obsidian hydration values were measurements that were greater than one standard deviation from the mean and in most cases were at least two standard deviations from the modal value. These anomalous (excessively small or large) readings were identified as outlying values and omitted from cluster sample statistics. The metrical data for each suite of readings includes mean, standard deviation, number of rim readings, and the coefficient of variation (cv). The latter measure, which is calculated by dividing the standard deviation by the mean, has been found useful in comparing multiple samples with varying means (Blalock 1979:84). The CV also provides a useful statistic to evaluate a sample's relative homogeneity. Single-period deposits have been defined as having a CV of 0.25 or less and having other chronological information (when available) largely consistent with that specific temporal period placement.

Obsidian hydration readings were rounded to the nearest 0.1 μ. Rim values in excess of 14.0 μ were dismissed from further consideration and are presumed, in most cases, to represent old, natural surfaces. The possibility exists that these very old hydration measurements might represent very early stoneworking. This alternative should not be ruled out.

Discussions with Rob Jackson and other specialists in obsidian hydration data interpretation have suggested that the early (1980–1985) hydration rim measurements produced by the laboratory of Joseph Michels are at odds with the findings of other researchers responsible for the bulk of the present hydration analysis. Michels' hydration rim measurement readings are consistently smaller than

Table 4.2 Chronological Periods for the Kern Plateau.

Bettinger and Taylor 1974		Gilreath and Hildebrandt 1997		Present Study		
Designation	Interval	Designation	Interval	Designation	Interval	Upland Coso Glass Hydration
Marana	650 BP–contact	Marana	650–200 BP	Chimney	650 BP–contact	<2.4 μ
Haiwee	1350–650 BP	Haiwee	1275–650 BP	Sawtooth	1350–650 BP	2.4–3.7 μ
Newberry	3150–1350 BP	Newberry	3500–1275 BP	Canebrake	3500–1350 BP	3.7–6.6 μ
Little Lake	6000–3150 BP	Little Lake	5500–3500 BP	Lamont	8500–3500 BP	6.6–10.1 μ
Mojave	pre-6000 BP	Early	pre-5500 BP	Kennedy	13500–8500 BP	10.1–13.9 μ

comparable readings on the same artifacts measured by all other laboratories and therefore are excluded from this study.

Double readings for flakes and points were uncommon (N = 11 or 2.3%). When they were present, the smaller of the two readings was used for statistical analysis and chronological placement. The smaller reading represents the most recent event. Dual rims are usually the result of scavenging of an older piece of obsidian from an earlier deposit and hence exhibit a larger rim along with a smaller one. The larger rim is often still intact since reworking of a stone tool often incorporates only portions of the original artifact leaving substantial portions of the original artifact form intact.

For many site loci, chronological data strongly suggest a single-period occupation, especially where hydration value clusters are largely consistent with the ages of time-sensitive artifacts (points, beads, or pottery) and radiocarbon assays. Often this correlation occurred for a single feature (e.g., a rock ring) or a site locus (e.g., a midden area), where the distribution of cultural remains was spatially segregated. Site features or areas appearing to have been used during two distinct and consecutive periods were so indicated by a bimodal distribution of hydration measurements. In those instances, cultural materials were assigned to both periods simultaneously and were so designated.

For the remainder of the site loci, a diverse range of hydration measures indicated multiple periods of activity. Such areas are considered temporally mixed. Other areas produced no chronological information and thus were classified as indeterminate. The chronological assessment of each component of the study sites is summarized in Table 4.3.

Obsidian Hydration Dating

Obsidian hydration studies are based on the principle that moisture penetrates volcanic glass at a predictable and quantifiable rate and hence the elapsed time since the glass was broken or artificially flaked can be calculated. Irving Friedman and Robert L. Smith (1960) first developed the obsidian hydration dating technique. They discovered that obsidian or volcanic glass absorbs small amounts of water over time. As soon as a fresh surface of a tool or flake is exposed to the elements it will begin to grow a rind. Research has conclusively shown that hydration rims on younger artifacts are smaller than those on older objects. In other words, the longer the time elapsed, the more water is absorbed. The

Table 4.3 Kern Plateau Site Components.

Period & Site Designation	N	Hydration values (in microns)	Mean	sd	cv	Classified Projectile Points**	^{14}C Assays	Other Temporal Data***
	Hydration Data*							
Chimney								
KER1273	1	2.0	2.0	0	NA			
KER-1276A	1	2.4	2.4	0	NA			
KER-1276B	1	2.2	2.2	0	NA	1 E	Modern	
KER-748	2	1.7, 1.9	1.8	0.14	0.07	1 DSN 1 CT 1 G		1 H2 shell bead
KER-743	1	1.6	1.6	0	NA	2 CT 1 HBN	Modern	1 stone bead
KER-742	No data							1 potsherd, 1 E2b shell bead
TUL-484A	No data					2 CT		
TUL-482	3	1.4, 1.8, 1.9	1.7	0.24	0.14	4 CT		3 potsherds, 3 glass beads
TUL-485	No data							1 stone bead
TUL-625	No data					1 CT		2 glass beads
TUL-623	3	1.8, 2.1, 2.3	2.1	0.25	0.11			
TUL-632	No data					1 CT		1 glass bead
TUL-879B								8 potsherds
TUL-894	3	2.0, 2.2, 2.8	2.3	0.42	0.18	1 DSN 1 CT		
TUL-909	No data					1 DSN		1 stone bead
TUL-891 Surface							225±80	11 potsherds
Chimney/Sawtooth								
KER 1298	7	2.3, 2.4, 2.5, 2.6, 2.6, 2.9, 3.3, (5.1)	2.6	0.31	0.11	1 RS	190±70	1 potsherd, 1 glass bead
KER-1278	2	3.2/3.5, 3.4	3.3	0.10	0.04	1 CT		
KER-1286	8	(1.1), (1.6), 2.5, 2.5, 2.6, 2.6, 3.0, 3.1, 3.2, 3.7, 4.1, (15.9)	3.0	0.56	0.19	2 CT	150±50 210±50 Modern	41 potsherds, 3 glass beads 1 K1 shell bead, 1 Phoenix button
KER-737	No data					1 CT 1 RS		
TUL-483	No data					1 DSN 4 CT 1 RS 1 HCB		5 potsherds, 1 H2 shell bead
TUL-481	2	(.5), 2.3, 2.8	2.5	0.37	0.12	4 DSN 2 CT 2 RS 1 HBN		13 potsherds, 10 glass beads

Table 4.3 Kern Plateau Site Components (continued).

Hydration Data*

Period & Site Designation	N	Hydration values (in microns)	Mean	sd	cv	Classified Projectile Points**	^{14}C Assays	Other Temporal Data*
Chimney/Sawtooth								
TUL-620	No data					3 RS 1 CT		7 stone beads
TUL-890	5	2.4, 2.4, 2.8, 3.4, 3.9	3.0	0.66	0.22	1 RS		60 potsherds
TUL-889	2	2.6 (3.9)	2.6	NA	NA	2 CT		
KER-1275	1	3.4/17.4	3.4	0	NA	1 HBN		62 potsherds
TUL-634	2	3.0, 3.7	3.3	0.5	0.15	1 CT		
Sawtooth								
KER-1296	18	1.5, 2.0, 2.2, 2.2, 2.7, 2.7, 3.0, 3.0, 3.0, 3.2, 3.3, 3.3, 3.4, 3.4, 3.4, 3.4, 3.5, 4.0	3.0	0.63	0.21	1 DSN		
KER-1299	7	2.3, 2.4, 2.5, 2.5, 3.2, 3.2, 4.0	2.9	0.62	0.21	2 RS		1 shell bead
KER-1284	3	2.8, 3.0, 3.5	3.1	0.36	0.11	1 RS		
KER-1285	4	3.4, 3.6, 3.6, 4.3	3.7	0.39	0.10			
TUL-484B	No data					1 RS		
TUL-484C	No data					1 RS		
TUL-619	5	(.7),(1.8), 2.2, 2.5, 3.4, 3.5, 4	3.1	0.75	0.24	1 DSN 1 HBN		
TUL-890B	6	2.4, 2.7, 2.7, 2.9, 3.2, 3.6, (5.5)	2.9	0.43	0.15	2 RS	1280±90	1 potsherd
TUL-890C	3	2.1, 3.2, 3.2	2.8	0.63	0.22			
TUL-767	No data					1 EG		
TUL-891E. Midden	4	2.7, 3.2, 3.3, 4.6	3.4	0.81	0.24			
TUL-891 W. Midden	7	2.5, 2.6, 2.9, 2.9, 3.0, 3.1, 3.1, (4.5)	2.9	0.24	0.08			
TUL-891 Depression	5	2.0, 2.0, 2.4/3, 2.9, 3.0	2.4	0.45	0.18			
TUL-625 Rock Rings (B)	2	2.4, 3.3, (5.0)	2.8	0.63	0.22			
Sawtooth/Canebrake								
KER-1269	4	(1.3), (1.4), (1.9), 3.6, 3.7, 4.1, 4.1	3.9	0.24	0.06	2 RS		
Canebrake								
TUL-630	7	3.4, 3.5, 3.6, 3.6, 4, 4.1, 4.5	3.8	0.43	0.11			

Table 4.3 Kern Plateau Site Components (continued).

Hydration Data*

Period & Site Designation	N	Hydration values (in microns)	Mean	sd	cv	Classified Projectile Points**	^{14}C Assays	Other Temporal Data***
Canebrake								
KER-1270	2	3.2, 4.5	3.8	0.9	0.23			
KER-1971	2	4.2, 5.0	4.6	0.57	0.12			
KER 1299E	2	5.5, 6.0	5.7	0.35	0.06			
KER-1279	9	3.2, 3.4, 3.5, 3.7, 3.7, 3.9, 4.5, 4.6, 5.5, (20.2)	4.0	0.73	0.18			
KER-1280	1	4.0	4.0	0	NA			
KER-1282	2	3.4, 8.6	6.2	1.4	0.22			
TUL-889B	1	4.5	4.5	0	NA			
TUL-895	1	4.6	4.6	0	NA	1 E		
TUL-887	2	(2.2), 4.3, 4.5, (8.2)	4.4	0.14	0.03	1 RS 1 E		
TUL-630	7	3.4, 3.5, 3.6, 3.6, 4, 4.1, 4.5	3.8	0.43	0.11			
Canebrake/Lamont								
TUL-889A	9	(2.6), 4.4, 4.7, 5.6, 5.9, 6.2, 6.3, 6.5, 7.4, 8.9	6.2	1.3	0.21	3 HCB 4 HBN		1 stone bead
TUL-889C	5	5.7, 6.3, 6.4, 7.4, 8.1/9.1	6.2	1.1	0.17	2 PT 1 RS		
Lamont								
TUL-616	...	(2.0)	...	...	...	1 PT		
Lamont/Kennedy								
TUL-897	5	(4.7), (5.5), (5.6), 6, 6.2, 8.5, 9/10.3, 11.4	8.3	2.2	0.25	1 PT 1 CB		
Multiple								
KER-1297	6	3.6, 4.1, 4.6, 6, 6.1, 6.4	5.1	1.2	0.23	2 RS 1 HBN		75 potsherds, 1 glass bead
KER-1298	29	1.2, 1.4, 1.4, 1.8, 1.9, 2.0, 2.3, 2.4, 2.5, 2.5, 2.6, 2.6, 2.8, 3.0, 3.0, 3.1, 3.4, 3.4, 3.5, 3.6, 3.7, 3.7, 4.2, 4.3, 4.6, 4.7, 4.7, 4.9, 5.2	3.1	1.1	0.35	8 CT 2 E 1 G 1 HCB	295±80 Modern	29 potsherds, 234 glass beads
KER-1277	7	(1.9), 2.6, 2.6, 3.6, 4.1,4.4, 4.6, 4.9	3.8	0.93	0.24	1 E	Modern 325±100	
KER-1281	11	1.8, 2, 2.1, 2.1, 2.9, 2.9, 3.0, 3.3, 3.6, 3.7, 4.0	2.8	0.76	0.27	1 DSN 2 CT 2 E		

Table 4.3 Kern Plateau Site Components (continued).

Hydration Data*

Period & Site Designation	N	Hydration values (in microns)	Mean	sd	cv	Classified Projectile Points**	^{14}C Assays	Other Temporal Data***
Multiple								
TUL488N	23	1.0, 1.1, 1.2, 1.8, 2.0, 2.0, 2.0, 2.0, 2.1, 2.1, 2.3, 2.5, 2.7, 3.0, 3.0, 3.7, 3.7, 3.9, 4.0, 4.4, 5.2, 6.6, 10.7	2.9	1	0.34	5 DSN 9 CT 8 RS 1 EG 2 HBN 1 PT		3 potsherds, 13 stone beads , 3 A1a, 3 E2b and 1 K1 shell beads
TUL-629	21	.8, 1.8, 2.0, 2.2, 2.4, 2.6, 3.0, 3.0, 3.0, 3.0, 3.1, 3.4, 3.6, 3.8, 4.4, 4.6, 4.8, 5.0, 5.1, 5.2, 6.1, 6.2	3.3	1.35	0.41	2 CT 2 RS 1 HCB		1 potsherd, 18 glass beads, 6 stone beads, 9 G1 and 1K1 shell beads
TUL-621	33	1.0, 1.0, 1.1, 1.2, 1.4, 1.4, 1.6, 1.6, 1.7, 2.0, 2.0, 2.0, 2.2, 2.3, 2.3, 2.3, 2.4, 2.4, 2.7, 2.7, 2.8, 3.0, 3.0, 3.0, 3.3, 3.5, 3.7, 4, 4, 4, 4.5, 4.7, 4.7	2.6	1.1	0.42	5 CT 4 DSN 4 RS		4 potsherds, 6 glass beads, 7 stone beads
TUL-617	No data					1 HBN		1 glass bead
KER-1283	7	1.1, 1.6, 2.9, 3.3, 3.6, 3.8, 4.0, (4.2), (4.4)	2.9	1.2	0.41	4 CT 1 RS		
TUL-636	7	2.3, 2.7, 3, 3.1, 3.2, 3.6, 3.7	3	0.52	0.17	2 CT 1 HBN 1 E		
TUL-877	10	(1.6), (1.6), 2.4, 3.2, 3.3, 3.5, 4.6, 4.6, 5.7, 5.7, 5.8, 6.6, (7.5)	4.4	1.4	0.31	3 RS 1 HBN	495 ± 165 590 ± 150 765 ± 170	1 glass bead
TUL-879A	27	1.1, 1.3, 1.4, 1.5, 1.7, 1.7, 1.7, 1.7, 1.8, 2.1 2.1, 2.5, 2.7, 2.8, 2.9, 3, 3, 3, 3, 3.2, 3.5, 3.6, 4.1, 4.9, 5.2, 6.9, 7.7	2.8	1.7	0.61	12 DSN 16 CT 2 RS 1 E 1 HCB	Modern 245 ± 75 250 ± 75 320 ±65 395 ±75 570 ±75 635 ±75 1110 ±160	11 potsherds, 4 stone beads, 2 E2b shell beads
TUL-882	No data					1 CT 1 DSN 1 E		
TUL-884	5	3.7/7.4, 3.9, 4.3, 4.7, 7.4	3.1	1.8	0.58	1 DSN		

Table 4.3 Kern Plateau Site Components (continued).

Hydration Data*

Period & Site Designation	N	Hydration values (in microns)	Mean	sd	cv	Classified Projectile Points**	^{14}C Assays	Other Temporal Data***
Multiple								
TUL-511	2	1.2, 8.0	4.6	5	1	2 DSN 1 E		70 potsherds, 2 stone beads
TUL-896	14	1.5, 1.9, 2.2, 3, 3.5, 3.9, 4, 4.2, 4.6, 5.2, 5.5, 5.6, 6.5, 8.9				1 DSN 4 E 2 HBN		
TUL-898	46	1.7, 1.9, 2.4, 2.6, 2.7, 2.9, 2.9, 2.9, 3.1, 3.2, 3.3, 3.2/3.8, 3.3, 3.3, 3.4/21.1, 3.4, 3.5, 3.5, 3.5, 3.5, 3.5, 3.6, 4.1, 4.3, 4.3, 4.4, 4.4, 4.4, 4.4, 4.9, 5.0, 5.2, 5.4, 5.6, 5.7, 5.9, 5.9/12.8, 6.1, 6.6, 7.2, 7.4, 7.7, 7.7, 7.8, 10.1, 10.6/12.1	4.6	2	0.4	1 DSN 1 HBN	820±80	1 potsherd, 3 stone beads
TUL-899	37	1.6, 2.1, 2.6/6.6, 2.6/4.9, 2.9, 3.3, 3.2, 3.3, 3.4, 3.4, 3.9, 4.1, 4.1, 4.4, 4.5, 4.9, 5, 5.1, 5.4, 5.5, 5.5, 5.6, 5.6, 5.7, 5.8, 6.4, 6.6, 7.1, 7.1, 7.6, 7.7, 7.7, 7.8, 9.1, 9.6, 12.5, 13.9	5.6	3	0.5	1 E 2 HBN 4 PT 1 CB		

The following sites contained no data with which to determine their age:

KER-1272	KER1274	TUL-487	TUL-488S	TUL-489
KER-744	KER-741	TUL-618	TUL-628	TUL-885
KER-747	KER-746	TUL-632	TUL-881	TUL-886
KER-738	KER-745	TUL-878	TUL-883	TUL-888
TUL-484D	TUL-480	TUL-880	TUL-889	KER-1299E Rock Rings

KEY:

**CB = Concave Base, CT = Cottonwood, DSN = Desert Side-notched, E = Elko, EG = Eastgate
G = Gypsum, HBN = Humboldt Basal-notched, HCB = Humboldt Concave Base,
RS = Rose Spring, PT = Pinto/Little Lake,
* ( ) Outlier values excluded from statistical calculations

KEY:

*** Bead types per Bennyhoff and Hughes 1987
A1 Small spire-lopped *Olivella* beads
E2b Thick-lipped *Olivella* beads
G1 Tiny saucer *Olivella* beads
H2 *Olivella* disks drilled w/ metal needles
K1 *Olivella* callus cup beads
Stone beads are serpentinite/talc disks

cv = Coefficient of Variation, sd = Standard Deviation,
N = Number of Hydration Samples

water forms a thin surface rind (hydration rim) which is visible under an optical microscope. These hydration rinds or rims are measured in microns. One micron equals 0.000039 of an inch.

Within the confines of the China Lake Naval Air Weapons Station, in the vicinity of Sugarloaf Mountain in the Coso Range of eastern California, lie many seams and outcrops of high-quality obsidian. Coso volcanic glass has been the focus of intensive studies and may be one of the "most thoroughly investigated obsidians in North America" (Gilreath and Hildebrandt 1997:10). These studies have spawned a plethora of alternative views on the proper hydration rate for dating Coso obsidian artifacts (Basgall 1990; Basgall and Hall 2000; Drews and Elston 1983; Ericson 1977, 1978a, 1978b; Garfinkel et al. 1980, 1984; Hildebrandt and Ruby 2003; King 2000; McGuire and Garfinkel 1980; McGuire et al. 1982; Meighan 1978; 1981; Pearson 1995; Rosenthal et al. 2001; Schiffman and Garfinkel 1981b).

Basgall's Coso Obsidian Hydration Rate

Mark Basgall (1990) introduced effective hydration temperature (EHT) into the Coso hydration equation and paired rims with associated radiocarbon dates to develop the most widely accepted curvilinear rate. This rate is derived from the extensive set of radiocarbon dates and Coso obsidian hydration rim values from the Lubkin Creek site, INY-30 (Basgall and McGuire 1988). Basgall and Hall (2000) and King (2000) have proposed some minor refinements to that rate. Basgall's 1990 hydration rate continues to be the most widely accepted formula, as it factors in mean annual temperature in the area from which the archaeological remains were recovered, and does explain much of the variability in the Coso hydration measurements. The rate he developed is

$$\text{Years B.P.} = 31.622\ X^{2.32},$$

where years B.P. is radiocarbon years before present (A.D. 1950) and X is the hydration measurement in microns (Basgall 1990:7).

This rate was used to advantage by Gilreath and Hildebrandt (1997) for their study of lowland sites within the Coso Volcanic Field in the southwest Great Basin. In reviewing Coso data from the Kern Plateau, Gilreath and Hildebrandt applied the Basgall rate with an effective hydration temperature (EHT) correction factor based on climatic data from the Grant Grove area. A number of researchers (Delacorte 1999; Gilreath and Hildebrandt 1997; Rosenthal et al. 2001) have cautioned, and the rate developer himself (Basgall 1993:85) agrees, that Basgall's Coso hydration equation overestimates the age of some materials. The rate particularly misrespresents the age of Early Holocene artifacts with hydration rims thicker than 10.0 or so microns (Table 4.4). One of the reasons for this problem is that hydration/radiocarbon age pairings of this age are rare and so secure dates for such ancient specimens are difficult to establish. Additionally, these associated radiocarbon dates are not routinely calibrated, for that reason most obsidian hydration rates underestimate the true age of the artifacts they date by as much as 1,000 to 2,000 years (Fiedel 1999).

Further complicating the issue, the EHT correction factor used with the Basgall Coso hydration rate may itself be problematic (Hildebrandt and Ruby 2000; Jones and Waugh 1995). Hildebrandt and Ruby (2000) recently reported that the Basgall formula, incorporating the characteristic EHT correction factor, consistently underestimated the age of many time-sensitive point types, especially those more ancient than several thousand years, from the high-elevation piñon forests in the Coso Range. Other factors affect the EHT and cause it to fluctuate through time. Among these are paleoclimatic change, pedoturbation (buried versus surface contexts), site aspect, vegetation cover, latitude, and elevation.

Consequently, EHTs are estimates influenced by a wide variety of environmental factors. The preferred strategy is to develop obsidian hydration and radiocarbon pairs for each locality and period since a single formula rarely provides reasonable age estimates for all times and places. For the present study such data were absent; consequently, alternative methods were used to develop an empirical hydration rate equation that might better approximate the age of prehistoric cultural materials in the high-elevation piñon forests of the Kern Plateau and vicinity.

Pearson's Coso Obsidian Hydration Rate

The inability to date early Holocene materials is not a limitation of the Pearson rate. Pearson's alternative obsidian hydration dating approach provides a set of reasonable and relatively accurate dates for the full span of human occupation in the southern Owens Valley (Pearson 1995). Analyzing time-sensitive point types and the distribution of Coso obsidian hydration readings, Pearson identified key benchmarks or transition points between the types. Using data from INY-30 (the Lubkin Creek Site), INY-372 (Rose Spring), and several sites at Little Lake (Stahl Site, Stahl Site Cave, and the *Pagunda* Village at the edge of Little Lake), he evaluated 401 Coso obsidian hydration rim measurements on projectile points and flakes with 44 associated radiocarbon dates to determine a local chronology for the Little Lake area.

Pearson correlated hydration measurements with beginning and ending dates for time-sensitive projectile point types. He then suggested that certain measurements represented particular calendar dates. To independently evaluate these relationships, he compared those benchmarks with associated radiocarbon dates and hydration measurements for these point types in the Little Lake vicinity. The rate Pearson developed is also a curvilinear equation converting hydration measurements into age in years: *y = number of microns (125+ [number of microns X 25]),* where y equals radiocarbon years before present (Table 4.4). That rate is intended to apply only to sites in the vicinity of Little Lake in the southern Owens Valley where EHT would be roughly equivalent.

Proposed Kern Plateau Coso Hydration Rate

In order to develop an appropriate rate that factors in the different sizes of rims and presumably slower hydration rate and differing EHT for the higher elevations (from 5000 to 7000 feet) on the adjacent Kern Plateau, a method comparable to that applied by Pearson was used. Based on an examination of the rim readings and correlating hydration measurements with the beginning, midpoint and ending dates of the time-sensitive projectile point types, a rate was developed and is presented here. That rate is also a curvilinear equation transforming hydration rims into an age in years: *y = number of microns (150+[number of microns X 60]),* where y equals radiocarbon years before present (A.D. 1950).

The resulting temporal periods are based on the above described equation and accommodate the means, standard deviations, and ranges of rim readings for the various point forms. The equation predicts the age ranges of the points reasonably well. The rate produces dates close to generally accepted radiocarbon ages (Table 4.5). This proposed rate also predicts the appropriate chronological period for time-sensitive projectile point forms with greater accuracy than the Basgall Coso rate even with an appropriate EHT factor incorporated into that formula (Tables 4.4 and 4.9).

From inspection of the Kern Plateau obsidian hydration data (Table 4.7 and 4.8), it appears that during the hydration process each micron of hydration takes approximately 60 years longer to form than the preceding micron and that the first micron of Coso hydration represents approximately 200 years. The smallest rim reading for a Desert Series point is 0.5 μ and that rim would equate, using the proposed formula, to an age of 90 years, very close to the historic date of 100 B.P. (ca. A.D. 1850)

Table 4.4 Coso Obsidian Hydration Rate Chronology Comparison.

Lowland Coso Rates

Sequence	Hydration Values (in microns)	Time Span* (Years B.P.)	Basgall Rate	Pearson Rate
Marana	<3.7	< 650	658	805
Haiwee	3.7–4.9	650–1,350	658–1,262	805–1,212
Newberry	4.9–7.6	1,350–3,500	1,262–3,495	1,212–2,394
Little Lake	7.6–16.0**	3,500–8,500**	3,495–19,658	2,394–8,400
Early	16.0–21.1	8,500–13,500	19,658–22,627	8,400–13,768

Upland Coso: Proposed Kern Plateau Rate

Sequence	Hydration Values (in microns)	Time Span* (Years B.P.)	Basgall Rate***	Proposed Rate
Chimney	<2.4	< 650	490	706
Sawtooth	2.4–3.7	650–1,350	490–1,341	706–1,376
Canebrake	3.7–6.6	1,350–3,500	1,341–4,963	1,376–3,603
Lamont	6.6–10.1	3,500–8,500**	4,963–13,784	3,603–8,474
Kennedy	10.1–13.9	8,500–13,500	13,784–28,932	8,474–13,678

KEY:
*After Gilreath (1999:12), with revisions by Rosenthal et al. (2001) and as discussed here.
**Ending date for Little Lake Period and terminus for Pinto Points revised per discussion herein.
***Basgall rate with Grant Grove EHT as presented in Gilreath and Hildebrandt (1997).

when protracted Euro-American contact is first recognized in the area. The largest hydration rim that is presumed to be cultural in origin is 13.9 μ, that measurement would convert to an age of 13,678 years — not an unreasonable age estimate for the very early use of the area.

Using the above formula, 2.4 μ represents the transition point between the Chimney and Sawtooth periods and denotes an age of 706 years — close to the generally accepted date of 650 B.P. that most researchers agree separates the most intensive periods of use for the Rose Spring from Desert Series points. Temporal bracketing of the Rose Spring and Eastgate points between 2.4 and 3.7 μ encompasses the majority (78%) of Coso hydration readings for these forms.

The largest readings on Rose Spring and Eastgate points are in the 3.7 μ range. That hydration measurement equates to a date of 1376 B.P. (with the present = AD 1950) or A.D. 574, very near the generally accepted initiation date for the Rosegate (Rose Spring and Eastgate combined) series at A.D. 600.

Elko and Humboldt series points largely date from the Canebrake Period (3500 to 1350 B.P.) and the bulk of our Kern Plateau samples had hydration rims ranging from 3.7 to 6.6 μ., rims that would be associated with that time span. A rim reading of 6.6 μ was chosen to mark the beginning of the Canebrake Period at 3500 B.P. equating with a rate-derived age of 3603 B.P. That date marks the generally accepted time of the initial appearance of Elko-series points (Bettinger and Taylor 1974; Heizer and Hester 1978; Justice 2002; Thomas 1981).

Bifurcate stemmed points (Pinto/Little Lake types) have a maximum hydration rim reading in our study sample of 10.7 μ. Basgall and Hall (2000:266) suggest that those point types date no earlier than ca. 7500 B.P. However, this claim discounts a wide array of earlier radiocarbon dates. Such dates have been obtained from Floodpond, Rogers Ridge, and Awl sites — all firmly associated with Pinto-age materials. The dates from the sites strongly indicate an antiquity greater than previously allowed by Basgall and Hall (2000) with an initial date of ca. 9000 B.P. (calibrated radiocarbon age). An even more ancient early Holocene age is possible but not confirmed.

Discussions with other researchers indicate that many prehistorians are now more inclined to accept these ancient middle Holocene dates for Pinto material and the early radiocarbon dates from the Stahl site (William Hildebrandt personal communication; Schroth 1994). Hydration rims of 7.6 to 16.0 μ on Pinto points at the Stahl site further support such a position. Those rims convert to dates from ca. 2400 to 8400 B.P. using the Pearson formula.

Using the Kern Plateau hydration rate would provide an approximation for the earliest of the Pinto/Little Lake points recovered from the study sites at ca. 8500 (calibrated) years B.P. Pinto or Little Lake points fall within the Lamont Period and exhibit a range of hydration rims on Coso obsidian running from 6.6 to 10.7 μ. The oldest and largest rim for a Pinto/Little Lake point coverts to an age of 8474 calibrated radiocarbon years ago. Such a date for the inception of Pinto/Little Lake points is reasonably consistent with our current understanding of the age of these points in the southwestern Great Basin.

Two concave base points assigned to the Kennedy Period have rims of 11.4 and 12.5 μ, equivalent to calendar dates of 9507 and 11,250 B.P. These two dates are ancient but not outside the bounds of the generally accepted age for aboriginal activity. The Kern Plateau data are consistent with independent estimates for the antiquity of these very early point forms and are supported by appropriately calibrated radiocarbon assays (cf. Fiedel 1999; Gilreath and Hildebrandt 1997; Justice 2002).

The reliability of Coso obsidian hydration data as a chronological index has been repeatedly shown by correlation of time-sensitive projectile point forms and hydration readings and by radiocarbon dates and associated hydration cluster values (Gilreath and Hildebrandt 1997). Nevertheless, hydration measurements are not amenable to great precision and yield only a general indication of age, not an absolute date. In the interest of accuracy, hydration rims are therefore not normally reported with specific dates. Given our reluctance to portray the sets of hydration rims with a greater accuracy level than is generally accepted, we used average rim readings associated with particular time periods dating site components (Table 4.3).

Radiocarbon Determinations

Twenty-six radiocarbon dates were obtained from 13 archaeological site components (Table 4.5). All dates were based on carbonized plant materials. The data in Table 4.5 suggest that several deposits maintain stratigraphic integrity. Age determinations fall within an expected mode from the oldest to

Table 4.5 Kern Plateau Radiocarbon Dates.

Laboratory Number	Uncorrected ^{14}C Age (yr B.P.)	Uncorrected ^{14}C Date	Uncorrected Range Of ^{14}C Date at 2 Sigmas	Calibrated Range of ^{14}C Date at 2 Sigmas	Locality	Site Name	Unit	Depth (cm d.b.s.)	Sample Substance	Reference
UCR 1258	295±80	A.D. 1655	A.D. 1575–1735	A.D. 1436–1693	Scodie Mts	KER-1298	Unit 4, Fea. 2	10 to 20	Charcoal	Ambro et al. 1981
UCR 1259	Modern	Modern	Modern	Not Applicable	Scodie Mts	KER-1298	Unit 4, Fea. 2	30 to 40	Charcoal	Ambro et al. 1981
UCR 1260	190±70	A.D. 1760	A.D. 1690–1830	A.D. 1629–1952	Scodie Mts	KER-1299	Unit 1	10 to 20	Charcoal	Ambro et al. 1981
UCR 1261	430±80	A.D. 1520	A.D. 1440–1600	A.D. 1394–1646	Scodie Mts	KER-1299	Unit 1	20 to 30	Charcoal	Ambro et al. 1981
UCR 1262	Modern	Modern	Modern	Not Applicable	Scodie Mts	KER-1276	Unit 1, Fea. 1	10 to 20	piñon hulls	Ambro et al. 1981
UCR 1263	Modern	Modern	Modern	Not Applicable	Scodie Mts	KER-1277	Unit 1	20 to 30	Charcoal	Ambro et al. 1981
UCR 1264	325±100	A.D. 1625	A.D. 1525–1725	A.D. 1411–1654	Scodie Mts	KER-1277	Unit 1	40 to 50	Charcoal	Ambro et al. 1981
UCR 1265	300±70	A.D. 1650	A.D. 1580–1720	A.D. 1442–1679	Scodie Mts	KER-1283	Unit 1	30 to 40	Charcoal	Ambro et al. 1981
UCR 1266	340±80	A.D. 1610	A.D. 1530–1690	A.D. 1420–1676	Scodie Mts	KER-1283	Unit 1	40 to 50	Charcoal	Ambro et al. 1981
IVC 17	Modern	Modern	Modern	Not Applicable	Morris Peak	KER-743	Rock Ring Hearth	10 to 20?	Charcoal	Garfinkel et al. 1980
I–13, 185	495±165	A.D. 1455	A.D. 1290–1620	A.D. 1187–1693	Rockhouse Bsn	TUL-877	N2E	20 to 30	Charcoal	Garfinkel et al. 1984
I–13, 186	590±150	A.D. 1360	A.D. 1210–1510	A.D. 1156–1648	Rockhouse Bsn	TUL-877	S38W19	30 to 40	Charcoal	Garfinkel et al. 1984
I–13, 187	765±170	A.D. 1185	A.D. 1015–1355	A.D. 938–1475	Rockhouse Bsn	TUL-877	S38W19	40 to 50	Charcoal	Garfinkel et al. 1984
I–13, 188	245±75	A.D. 1705	A.D. 1630–1780	A.D. 1477–1707	Rockhouse Bsn	TUL-879	N6W4	20 to 30	Charcoal	Garfinkel et al. 1984
I–13, 189	1110±160	A.D. 840	A.D. 680–1000	A.D. 646–1223	Rockhouse Bsn	TUL-879	N6W4	100 to 110	Charcoal	Garfinkel et al. 1984
I–13, 190	250±75	A.D. 1700	A.D. 1625–1775	A.D. 1471–1705	Rockhouse Bsn	TUL-879	N1W3	30 to 40	Charcoal	Garfinkel et al. 1984
I–13, 191	395±75	A.D. 1555	A.D. 1480–1630	A.D. 1413–1647	Rockhouse Bsn	TUL-879	N1W3	130 to 140	Charcoal	Garfinkel et al. 1984
I–13, 192	570±75	A.D. 1380	A.D. 1305–1455	A.D. 1283–1450	Rockhouse Bsn	TUL-879	N1W3	140 to 150	Charcoal	Garfinkel et al. 1984
UGa-3839	Modern	Modern	Modern	Not Applicable	Rockhouse Bsn	TUL-879	N2W2	10 to 20	Charcoal	McGuire 1981
UGa-3840	320±65	A.D. 1630	A.D. 1565–1695	A.D. 1442–1669	Rockhouse Bsn	TUL-879	N2W2	60 to 70	Charcoal	McGuire 1981
UGa-3841	635±85	A.D. 1315	A.D. 1230–1400	A.D. 1241–1438	Rockhouse Bsn	TUL-879	N2W2	100 to 110	Charcoal	McGuire 1981
UGa-3842	580±135	A.D. 1370	A.D. 1235–1505	A.D. 1187–1640	Rockhouse Bsn	TUL-890A	N1E1	10 to 20	Charcoal	McGuire 1981
UGa-3843	1280±90	A.D. 670	A.D. 580–760	A.D. 915–965	Rockhouse Bsn	TUL-890B	N4W4	40 to 50	Charcoal	McGuire 1981
UGa-3841	225±80	A.D. 1725	A.D. 1645–1805	A.D. 1609–1890	Rockhouse Bsn	TUL-891	Hearth Feat.	10 to 20	Charcoal	McGuire 1981
I–13, 665	820±80	A.D. 1130	A.D. 1050–1210	A.D. 1031–1297	Kennedy Mdws	TUL-898	N49/E7–8	30 to 40	Charcoal	Bard et al. 1985
Beta-4547	150±50	A.D. 1800	A.D. 1750–1850	A.D. 1664–1893	Scodie Mts	KER-1286	TU 2	10 to 20	Charcoal	McGuire 1983

youngest assay arrayed as a function of depth within their respective deposits. Yet obsidian tools or debitage confidently associated with the radiocarbon assays were uncommon.

Rim readings within the deposits showed that a great deal of "churning" of deposits had taken place. This churning is quite common in many archaeological sites, even those considered to have a fair degree of stratigraphic integrity. Most likely, cultural and natural agents including erosion, rodent activity, and aboriginal disturbances worked together to cause the movement of lightweight obsidian artifacts up and down in the deposit.

These factors generally preclude the development of a reasonable range of radiocarbon age-hydration rim pairings for the study sites. In almost every case the radiocarbon samples were aggregate collections of charcoal fragments from a single excavation level. As such, the assays are more of an average than a precise date. Many of the most recent radiocarbon dates come from sites in the Scodie Mountains. The two earliest dates came from Rockhouse Basin, and a slightly younger, but still rather old, date came from a Kennedy Meadows site (Table 4.5).

Projectile Points

In total, 222 projectile points complete enough to allow classification as to type were identified from surface collections and excavations (Table 4.1). In most cases, classification of these forms follows previous designations. Using the metric attributes defined by Thomas (1981), this study classified points into standard Great Basin types (Table 4.6). In some instances, prior classifications were determined to be in error, or more recent research led to differing conclusions regarding classification for certain forms (Basgall and Hall 2000; Gilreath and Hildebrandt 1997; Garfinkel and Yohe 2004).

Points were retrieved from 51 site loci. The projectile point inventory provides examples of most types commonly found throughout the Great Basin. These projectile points suggest that the study area was occupied from the terminal Pleistocene through the Chimney Period. These materials consist of 41 Desert Side-notched, 78 Cottonwood, 43 Rose Spring and two Eastgate, 19 Humboldt Basal-notched, eight Humboldt Concave Base, 19 Elko, two Gypsum, nine Little Lake or Pinto, and two Paleoindian Concave-based points. Looking only at their typological affiliation, 119 can be assigned to the Chimney Period (Desert Side-notched and Cottonwood), 45 to the Sawtooth time span (Rose Spring and Eastgate), 48 to the Canebrake interval (Elko, Gypsum, and Humboldt), and nine to the Lamont Period (Pinto). Two points are from the Kennedy Period (Concave Base).

Hydration rims on a sample of these points show that their mean values increase from the Desert series, to Rose Spring/Eastgate, to Humboldt Basal-notched, Humboldt Concave Base, and Elko, with the Little Lake/Pinto and Concave-based (Paleoindian) points exhibiting significantly larger readings (Table 4.7).

As shown in Tables 4.7 and 4.8, the hydration rims on the majority of the Desert series, Rosegate, and Humboldt Basal-notched point forms fall within the expected range for the various time periods identified.

Desert Side-notched Series (Figure 3.1 k)

Desert Side-notched points are small, triangular forms usually weighing less than 1.5 g with side notches placed high on their margins, (Baumhoff 1957; Baumhoff and Byrne 1959). Most such points recovered from the study sites also have indentations on their bases and would conform to the Sierra subtype. Two

Table 4.6 Summary of Metric Data for Projectile Points.

Desert Side-notched (n = 41)	ML	AL	MW	BW	TH	DSA	PSA	WT
N	18	17	30	30	34	37	37	16
Mean	19.8	19.2	11.9	11.9	3.0	207.8	171.3	0.5
s.d.	4.0	3.9	3.2	3.2	0.6	20.3	17.6	0.3
max.	27.6	27.3	24.5	24.5	4.0	200.0	200.0	1.0
min.	13.5	13	7.8	7.8	2.1	125.0	125.0	0.2

Cottonwood (n = 78)	ML	AL	MW	BW	TH	DSA	PSA	WT
N	48	48	71	71	78	NA	NA	39
Mean	22.1	20.7	11.6	11.5	3.1	NA	NA	0.6
s.d.	5.1	5.3	2.4	2.4	0.8	NA	NA	0.5
max.	37.6	37.6	16.8	16.8	6.0	NA	NA	2.3
min.	12.0	11.2	6.0	6.0	2.0	NA	NA	0.1

Rose Spring (n = 43)	ML	AL	MW	BW	TH	DSA	PSA	WT
N	23	24	37	41	43	42	43	17
Mean	23.0	24.0	37.0	41.0	43.0	42.0	43.0	17.0
s.d.	26.4	25.9	13.3	8.1	3.4	19.3	23.5	1.1
max.	38.7	38.7	17.5	13.1	4.4	185	106	1.8
min.	4.9	5.2	2.1	2.2	0.8	220	170	0.4

Eastgate (n = 2)	ML	AL	MW	BW	TH	DSA	PSA	WT
N	2	2	2	2	1	2	2	2
Mean	31.1	31.1	18.7	10.8	3.9	110.0	125.0	1.6
s.d.	9.3	9.3	0.3	6.7	0.0	14.1	49.5	0.4
max.	37.7	37.7	18.9	18.5	3.9	120.0	160.0	1.9
min.	24.5	24.5	18.5	6.9	3.9	100.0	90.0	1.3

Elko (n = 15) (Elko Corner-notched & Elko Eared)	ML	AL	MW	BW	TH	DSA	PSA	WT
N	5	5	12	12	14	13	10	1
Mean	27.9	25.8	25.2	23.2	6.1	183.7	149.1	7.0
s.d.	14.8	13.6	5.8	6.4	2.0	22.9	25.8	0.0
max.	45.3	43.3	33.7	32.0	12.2	220.0	220.0	7.0
min.	11.4	11.4	12.0	12.0	4.0	140.0	115.0	7.0

Elko (n = 4) (Elko Side-notched)	ML	AL	MW	BW	TH	DSA	PSA	WT
N	3	3	3	3	4	4	4	1
Mean	29.3	28.0	22.5	19.8	6.5	202.5	151.2	6.8
s.d.	5.5	4.6	3.3	5.3	0.7	17.1	16.5	0.0
max.	32.0	32.0	25.5	25.5	7.0	220.0	170.0	6.8
min.	23.0	23.0	19.0	15.0	5.5	180.0	130.0	6.8

Table 4.6 Summary of Metric Data for Projectile Points (continued).

Gypsum (n = 2)	ML	AL	MW	BW	TH	DSA	PSA	WT
N	0	0	1	1	1	1	1	0
Mean	0.0	0.0	27.0	16.0	6.0	180.0	85.0	0.0
s.d.	0.0	0.0	0.0	0.0	0.0	0.0	0.0	0.0
max.	0.0	0.0	27.0	16.0	6.0	180.0	85.0	0.0
min.	0.0	0.0	27.0	16.0	6.0	180.0	85.0	0.0

Humboldt Basal-notched (n = 19)	ML	AL	MW	BW	TH	DSA	PSA	WT
N	6	7	14	13	19	NA	NA	0
Mean	37.1	28.1	21.6	19.6	6.8	NA	NA	0.0
s.d.	9.8	8.3	4.3	4.6	1.3	NA	NA	0.0
max.	52.5	41.2	29.7	29.7	9.2	NA	NA	0.0
min.	28.0	20.0	16.4	14	4.1	NA	NA	0.0

Humboldt Concave Base (n = 8)	ML	AL	MW	BW	TH	DSA	PSA	WT
N	1	2	7	7	8	NA	NA	0
Mean	31.4	29.7	21.8	16.7	6.9	NA	NA	0.0
s.d.	0.0	0.4	6.1	4.4	2.6	NA	NA	0.0
max.	31.4	30.0	29.0	20.0	10.7	NA	NA	0.0
min.	31.4	29.4	15.8	11.4	4.5	NA	NA	0.0

Little Lake (Pinto) (n = 9)	ML	AL	MW	BW	TH	DSA	PSA	WT
N	8	8	9	8	8	5	5	3
Mean	26.6	25.1	23.1	20.3	7.6	196	114	4.9
s.d.	5.8	5.3	4.4	4.0	1.3	11.4	28.8	1.0
max.	30.5	28.9	31.2	26.5	9.0	210.0	150.0	5.5
min.	12.6	12.6	18.0	13.5	5.5	180.0	90.0	3.8

Concave Base (n = 2)	ML	AL	MW	BW	TH	DSA	PSA	WT
N	2	2	2	2	2	NA	NA	0
Mean	35.3	29.7	30.4	27.6	9.2	NA	NA	0.0
s.d.	10.9	4.6	9.3	5.4	1.1	NA	NA	0.0
max.	43.0	33.0	37.0	31.5	10.0	NA	NA	0.0
min.	27.5	26.5	23.8	23.8	8.5	NA	NA	0.0

KEY: AL axial length; BW basal width; DSA distal shoulder angle; NA not applicable; ML maximum length; MW maximum width; PSA proximal shoulder angle; TH thickness; WT weight. All measurements in millimeters.

points, formerly misidentified, do not have true side notches but instead have rather incurvate lateral margins and lack basal indentations (Garfinkel et al.1980:57, Figure 1 a and d). Both were retrieved from TUL-488N and represent a type of point not previously identified as a temporal diagnostic. Hydration rims for both specimens (4.4 and 5.2 μ) indicate that their age is commensurate with the Canebrake Period and comparable with Elko series points.

Otherwise, classification of the Desert Side-notched specimens was a straightforward and simple task, with 41 Desert Side-notched points recovered from 16 sites. With the exception of two outliers, most likely reworked points made from an older piece (2.8 and 4.7 μ), hydration readings on 12 of these

Table 4.7 Hydration Data Summary of Coso Obsidian Projectile Points from the Kern Plateau.

	N	mean	s.d.	Rim Values
Desert Side-notched	13	1.8	0.4	1.2, 1.4, 1.7, 1.8, 1.9, 2.0, 2.0, 2.0 2.1, 2.2, 2.3, 2.3, 2.8, (4.7)
Cottonwood	25	1.9	0.6	0.5, 1.0, 1.1, 1.4, 1.4, 1.6, 1.6, 1.6, 1.6 1.6, 1.7, 1.8, 1.8, 1.8, 1.9, 1.9, 2.0, 2.0 2.1, 2.4, 2.4, 2.6, 2.7, 3.0, 3.1, (3.6), (3.7)
Rose Spring/Eastgate	21	3.0	0.5	(1.0), (1.8), 2.2, 2.4, 2.4, 2.4, 2.4, 2.5 2.5, 2.8, 2.8, 2.9, 2.9, 3.0, 3.0, 3.0 3.1, 3.3, 3.6, 3.7, 3.7, 3.9, 4.0
Humboldt Basal-notched	13	5.7	1.4	3.7, 3.8, 4.6, 4.6, 5.2, 5.5, 5.7, 6.2, 6.3, 6.4, 6.5, 6.6, 8.9
Humboldt Concave Base	2	5.1	0.7	4.6, 5.6
Elko (Corner-notched and Eared)	6	6.3	2.4	(2.2), 3.7, 4.6, 4.9, 6.0, 8.9, 9.6, (12.5)
Elko Side-notched	2	5.4	0.8	4.9, 6.0
Pinto or Little Lake	4	8.6	1.7	(1.6), (3.9), 6.6, 8.1/9.1, 9.0, 10.7
Concave Base	2	11.9	0.8	11.4, 12.5
Total	88			

NOTE: ( ) — Value not included in calculations; / double band — smaller reading used.

Table 4.8 Frequency of Coso Obsidian Projectile Points by Type and Period.

		Point Types							
Period	Hyd. Range	Desert	Rose Spring	HBN	HCB	Elko	Pinto	Concave	Total
Chimney	<2.4	35	3	-	-	1	1	-	40
Sawtooth	2.4–3.7	5	18	-	-	1	1	-	25
Canebrake	3.7–6.6	1	2	12	2	5	-	-	22
Lamont	6.6–10.7	-	-	1	-	2	4	-	7
Kennedy	>10.7	-	-	-	-	1	-	2	3
Total		41	23	13	2	10	6	2	97

points conform closely to expectations of 2.4 μ or less for the Chimney Period. The age of Desert Series points is well established based on associated radiocarbon dates and other evidence. Most archaeologists agree that they date to the interval after 650 B.P. (Bettinger and Taylor 1974; Gilreath and Hildebrandt 1997; Thomas 1981) and as such are a hallmark of the Chimney Period. Yet, these same points become increasingly younger in age as one moves north and east out of the southern Owens Valley. Delacorte (1995) first recognized this trend and argues that these points may, in fact, be distinctive marker artifacts of Numic groups indicating their spread and population movement from a homeland in the Owens Valley less than a thousand years ago.

Table 4.9 Percentage of Chronologically Diagnostic Coso Obsidian Points Attributed to Correct Chronological Period.

Period	Chimney	Sawtooth	Canebrake	Lamont	Kennedy	Average
Rate						
A	80	78	40	17	0.0	43.0
B	77	78	76	66	100	79.4

KEY: A = Basgall formula with EHT conversion for Grant Grove as used by Gilreath and Hildebrandt (1997)
B = Hydration conversion formula as proposed in this study

Chimney Period (650–100 B.P.) Desert series includes Desert Side-notched and Cottonwood Triangular or Leaf-shaped (38 specimens).

Sawtooth Period (1350–650 B.P.) Rosegate series includes Rose Spring and Eastgate (21 specimens).

Canebrake Period (3500–1350 B.P.) Elko series includes all Elko Eared, Elko Corner-notched, and Elko Side-notched (10 specimens); Humboldt series includes Humboldt Basal-notched (13 specimens), and Humboldt Concave Base (2 specimens).

Lamont Period (8500–3500 B.P.) Little Lake/Pinto series (6 specimens).

Kennedy Period (13500–8500 B.P.) Concave Base type (2 specimens).

Cottonwood Series (Figure 3.1 j)

Cottonwood points are small (usually weighing <1.5 g) triangular points lacking notches (Riddell 1951), with margins typically straight to slightly concave and bases that are straight to deeply concave or notched. Examples having markedly convex bases and blades are referred to as Cottonwood Leaf-shaped points (Heizer and Baumhoff 1961; Lanning 1963; Thomas 1981). Both forms are recognized in the present collection. Most evidence, including radiocarbon assays and other indicators, suggest that in eastern California they are contemporary with Desert Side-notched points (650 B.P.–post contact), making them a second time marker of the Chimney Period. Seventy-eight Cottonwood points were identified from 18 sites. Almost one-third of these were retrieved from two site components with deep midden deposits (TUL-879 and TUL-488N) which contained almost as many Desert Side-notched points.

Hydration readings from 25 Cottonwood points mostly (75%) conformed to our expectations and fall within the suggested hydration range of the Chimney Period. Yet, six outliers are larger than expected (>2.4 µ) and are readings most likely derived from older, scavenged and reworked points.

Rose Spring Series (Figure 3.1 i)

Forty-three Rose Spring series points were collected from 16 sites. Rose Spring points were originally recognized from the type-site of that same name located in southern Owens Valley, at the edge of the Coso Range (Lanning 1963). The Rose Spring type is a small, narrow, triangular arrow point with a variety of stem forms. Rose Spring points are time markers for the interval from ca. 1350–650 B.P. (the Sawtooth Period) in the far southern Sierra and southwestern Great Basin (Basgall and McGuire 1988; Bettinger and Taylor 1974; Gilreath and Hildebrandt 1997; Thomas 1981; Yohe 1992). Hydration readings from 21 Rose Spring points, largely (78%) conform to our predictions and all but five specimens fall within the suggested hydration range of the Sawtooth Period (2.4–3.7 µ).

Eastgate Series (Figure 3.1 h)

First recognized at Wagon Jack Shelter (NV-Ch-119) in central Nevada (Heizer and Baumhoff 1961), Eastgate points have often been joined with the Rose Spring type into a group known as the Rosegate series, as defined by Thomas (1981). This name highlights the contemporaneity of Eastgate and Rose Spring points from ca. 1350–650 B.P. (Bettinger and Taylor 1974; Hester 1973; Thomas 1981). However, regional distributions of the two forms differ. Rose Spring points are far more common in the study area and generally within the southwestern Great Basin.

Eastgate points are morphologically distinct from Rose Spring points in that they have a wide triangular blade element with deep notching of the base, leaving squared or rounded shoulder barbs and sometimes an expanding stem. The Eastgate forms are also distinctive in that they have prominently barbed shoulders. These points also have blade forms that in some instances are slightly concave in outline. Many are quite large and broad having an outline similar to an equilateral triangle. The notches are narrow and completed in a fashion such that the point outline is uninterrupted. Hence, the barbs might be described as hanging or extending to the level of the base or even farther (Delacorte 1990:118). Eastgate points generally differ from Rose Spring forms in that they are more finely finished. Only two specimens were recovered at the study sites. A single hydration reading of 3.9 µ is available for an Eastgate point from TUL-488 and this measure conforms to the expected range of rim values at the early end of the Sawtooth interval.

Humboldt Series (Figure 3.1 d and e)

Heizer and Clewlow (1968) originally proposed the Humboldt types based on archaeological materials from the surface of the Humboldt Lakebed Site (NV-Ch-15) in western Nevada. These points are unshouldered, lanceolate forms with slight basal concavities to deep basal notches. Three variants of Humboldt series points were described initially: Concave Base A, Concave Base B, and Basal-notched. Most researchers subsequently merged the first two types as simply Concave Base (Heizer and Hester 1978). In the southwestern Great Basin, stratigraphic contexts and obsidian hydration readings argue for chronological placement of Humboldt Concave Base points (Figure 3.1 d) roughly synchronous with the Elko series and the most recent span of Pinto points, placing them in a time span from ca. 4000 to 1350 B.P. (Basgall and McGuire 1988; Delacorte 1999; Delacorte and McGuire 1993; Gilreath and Hildebrandt 1997; Hall 1983; Hall and Jackson 1989; Jackson 1985).

Eight Humboldt Concave Base points were recovered from four sites. Two hydration readings (4.6 and 5.6 µ) confirm placement in the Canebrake Period (3500–1350 B.P), contemporary with the Elko series and the majority of Humboldt Basal-notched points recovered from sites in the study area.

Morphological confusion occurs between the Humboldt Basal-notched form (Figure 3.1 e), the lookalike types of the Pinto Shoulderless (Harrington 1957) and Sierra Concave Base (Moratto 1972) forms. The former type is part of the Little Lake series and dates to the Lamont Period in the local sequence. It is very difficult to differentiate small proximal basal fragments of Pinto Shoulderless points from the Humboldt Basal-notched type. Also, researchers differ in their views as to whether the Pinto Shoulderless form should be combined with the Humboldt Basal-notched type (Delacorte et al. 1995: 68; Pearson 1995; and Schroth 1994 would like to combine the forms, and Basgall and Giambastiani 1995, among others, favor a split).

Based on the hydration readings from 13 Humboldt Basal-notched bifaces, all but one conforms to the hydration range characteristic of the Canebrake Period (3500–1350 B.P.). With respect to the Sierra

Concave Base and Humboldt Basal-notched confusion, it is rather difficult to make that distinction based on the small size of the present sample (13) and the similarity in the morphologies of these two forms (Stevens 2001). Nevertheless, the lanceolate, basal-notched, unshouldered points in this part of the far southern Sierra appear to have an initial date some 500 years or so earlier than that usually given for the Sierra Concave Base points farther north in the Sierra Nevada (Justice 2002; Moratto 1972). Additionally, obsidian hydration data for Sierra Concave Base forms from the southern Sierra foothills indicates that this type may have a lengthier duration than Humboldt Basal-notched forms (Stevens 2001). The Humboldt Basal-notched form appears to abruptly terminate during its peak period of popularity (ca. A.D. 800) while the Sierra Concave Base type continues until the later prehistoric era, perhaps as recently as ca. A.D. 1300 (Garfinkel and Yohe 2004; Stevens 2001).

A recent comprehensive review of the typological and chronological parameters of the Humboldt Basal-notched form in the southwestern Great Basin led researchers to identify wide- and narrow-based subtypes (Garfinkel and Yohe 2004). It was suggested that the former was more recent, dating from 1150–2450 B.P., while the latter was most popular during an earlier interval, from 2450–5950 B.P. The break point for the two variants would equate with a rim value of 5.3 μ based on the Kern Plateau hydration rate. Examination of the small sample of bifaces (10) from the study sites does not allow us to confidently differentiate between subtypes (Table 4.10).

Hydration data were obtained for 13 Humboldt Basal-notched bifaces recovered from eight sites. Excluding one outlying value (8.9 μ), their 12 obsidian hydration measurements range from 3.7 to 6.6 μ, indicating that most of these specimens are of an age equivalent to Elko and Gypsum points (Canebrake Period markers) with perhaps some limited persistence into the earliest portion of the Sawtooth interval. Many eastern California prehistorians corroborate this pattern and agree with the dating of these forms (Basgall and McGuire 1988; Gilreath and Hildebrandt 1997; Hall and Jackson 1989).

Elko Series (Figure 3.1 f)

Heizer and Baumhoff (1961) were the first to define Elko points. This series is composed of large, heavy, notched points with variable stem characteristics (Heizer et al. 1968; O'Connell 1967). The present sample includes eared, corner- and side-notched specimens. Contracting stem forms are assigned to the Gypsum type (see below). Elko series points are time markers for the Canebrake Period in the far southern high Sierra.

Table 4.10 Hydration Measurements for Kern Plateau Lanceolate Basal-notched Bifaces of Coso Obsidian.

Basal Width/Maximum Width Minimum* Measures in Millimeters	24 +	<24	Total
Hydration Rim Readings			
>5.3	0	6	6
3.7 to 5.3	2	2	4
Totals	2	8	10

* Only complete measurements tallied from the smaller of the two measures; either basal width or maximum width measurement metrics (whichever complete measure is available) are included; measures are all in millimeters. The measures included are only for the Coso obsidian specimens chemically characterized to source.

In the western Great Basin, Elko points consistently occur in contexts dating from 3500–1350 B.P. (Basgall and McGuire 1988; Bettinger and Taylor 1974; Gilreath and Hildebrandt 1997; Heizer and Hester 1978; Justice 2002; Thomas 1981). Such a chronological position is supported by a plethora of radiocarbon, stratigraphic and hydration data, although it is becoming increasingly apparent that large corner-notched and side-notched forms also occur in earlier contexts. Gilreath and Hildebrandt (1997) noted that more robust Elko points, especially those thicker than 6.5 mm, regularly produced hydration rinds that are more ancient than those of Canebrake Period artifacts. One explanation for this problem is the difficulty in distinguishing between the earlier Pinto and the more recent look alike Elko forms (Basgall and Hall 2000; Vaughan and Warren 1987).

Nineteen Elko-series points were recovered from 12 sites in the study area. Hydration measurements of eight specimens indicate that Elko forms in the far southern Sierra date largely to the Canebrake Period. Earlier examples of large corner-notched forms dating from the Lamont (two readings at 8.9 and 9.6 µ) and Kennedy (one reading at 12.5 µ) periods support the notion that this form has greater longevity than previously recognized. Thickness measurements on the Kern Plateau points do not seem to support the temporal/morphological distinctions identified by Gilreath and Hildebrandt (1997), having no evident correlation by age and thickness, but our sample is too small to make a confident determination on this matter.

Gypsum Series (Figure 3.1 g)

Large contracting–stem points were originally identified by Mark Raymond Harrington (1933) at Gypsum Cave in southern Nevada. Morphologically similar forms have been identified as Elko contracting-stem types (Heizer and Baumhoff 1961; Heizer et al. 1968; O'Connell 1967). Thomas (1981) calls similar points from Central Nevada Gatecliff Contracting-stem. Apart from their designations, these different terminologies also reflect apparent differences in chronology. The Gatecliff series attribution is meant to characterize a group of earlier, pre-Elko points, with dates ranging from 4950 to 3150 B.P. (Thomas 1981).

Nevertheless, evidence from the southwestern Great Basin consistently shows that forms are fully synchronous with Elko series points dating from 3500–1350 B.P. (Basgall and Giambastini 1995; Bettinger and Taylor 1974; Hall 1983; Hall and Jackson 1989). Two specimens were identified at two study area sites but neither specimen was measured using hydration analysis.

Pinto (Little Lake) Series (Figure 3.1 c)

Large, bifurcate-stemmed points have often been identified as members of the Pinto or Little Lake series (cf. Bettinger and Taylor 1974; Lanning 1963). Researchers have conjectured that the Pinto-like points from the Stahl site near Little Lake (Harrington 1957) were morphologically distinct from Pinto points named after the type locality in the Pinto Basin of the Mojave Desert (Amsden 1937; Campbell and Amsden 1934; Campbell and Campbell 1935; Schroth 1994).

Recent research (Basgall and Hall 2000) is counterintuitive in suggesting that the points from the Stahl site are largely indistinguishable from Mojave Desert examples. The research of Basgall and Hall further indicates that these robust forms have an age from 7500–4000 B.P., while gracile forms, more characteristics of the northern Great Basin, are equivalent to the Gatecliff Split-stem type previously identified by Thomas (1981). Those latter artifacts date to a more recent time, consistent with a range of 5000–3200 B.P. It has also become apparent that in the general region of eastern California considerable spatial overlap exists between the robust and gracile variants.

Nine Pinto/Little Lake series points were recognized at five sites. Hydration data are available for six specimens. Two anomalous obsidian hydration measures (1.6 and 3.9 μ) are probably due to scavenging and reworking of these points during later periods. The four remaining measures indicate ages commensurate with the Lamont Period (6.6, 8.1/9.1, 9.0, and 10.7 μ) and would yield dates ranging from 3604 to 8474 B.P. based on the Kern Plateau Coso obsidian hydration rate formula.

Concave Base (Figure 3.1 b)

Finally, it has become evident that a class of large, lanceolate, concave base projectiles exists that date to the late Pleistocene/early Holocene eras. These bifaces are very similar to the Humboldt series, yet they are often basally ground and sometimes exhibit grinding along their lateral margins. Concave base forms have been suggested to date ca.11,000–13,500 B.P. (Fiedel 1999; Pendleton 1979). Similar forms are recognized at El Portal on the Merced River (Hull and Moratto 1999), in Long Valley (Basgall 1988), in the Coso Range (Gilreath and Hildebrandt 1997), at the Sherwin Summit (Eerkens and King 2002), in Bridgeport Valley (Halford 2001), in the Black Rock desert (Clewlow 1968), at Tulare Lake (Riddell and Olson 1969), and in the Tonopah area of Nevada (Tuohy 1984). Two points from two sites in Kennedy Meadows (TUL-897 and TUL-899) are assigned to this category. Obsidian hydration rims on these two points (11.4 and 12.5 μ) would equate to 9507 and 11,250 calendar years B.P., roughly commensurate with prior estimates for the age of these forms (Fiedel 1999; Justice 2002).

Ceramics

Some 396 sherds of plain brownware ceramics (Riddell 1951; Riddell and Riddell 1956; Steward 1928) were recovered from 20 sites. Previous reviews of these materials have confirmed that this pottery is properly classified as a type of Owens Valley Brownware (Griset 1981; May 1980, 1981). Griset (1981) synthesized chronometric information for the Kern Plateau and argued that pottery was introduced there sometime after 450 B.P. (ca. A.D. 1500). Research from east of the Sierra also indicates that pottery came into widespread use there only during the last five centuries (Basgall and Giambastini 1995; Basgall and McGuire 1988; Delacorte 1999). This would place Owens Valley Brownware as a consistent hallmark of Chimney Period components in the study area and vicinity.

Olivella Shell Beads

Twenty-four *Olivella* shell beads were recovered from seven study sites. Prior research provided sufficient information to place all but one into types provided by Bennyhoff and Hughes (1987) (Table 4.11, Figure 4.1). *Olivella* disks drilled with metal needles (H series) are Mission Period (A.D. 1770–1834) forms, according to Bennyhoff and Hughes (1987:135) and King (1990:179–182). Callus cup beads (K series) were manufactured anytime from about A.D. 1150 to 1770 based on dates and cultural sequences from southern California (King 1990:166). Finally, Thick-lipped forms (E2 category) are common during Phase 2b of the Late Period in central California or the period from A.D. 1700 to 1800 (Bennyhoff and Hughes 1987:127–129).

Other forms are less diagnostic. Nine G1 (Tiny Saucer) beads occur in a wide range of temporal contexts (Bennyhoff and Hughes 1987). Three A1a (Small Spire-lopped) beads are also poor time markers (Bennyhoff and Hughes 1987). They are most common during the Early Period in central California (3000–500 B.C.); but are popular again during Phase 1 of the Late Period (A.D. 900–1500) (Bennyhoff and Hughes 1987). The latter bead form also occurs in varying quantities during the Middle Period from 200 B.C. to A.D. 700.

Table 4.11 *Olivella* Shell Beads From Kern Pl

	A1a	E2b	G1	H2		
KER-1286	-	-	-	-		
KER-748	-	-	-	1	-	
KER-742	-	1	-	-	-	1
TUL-483	-	-	-	1	-	1
TUL-488N	3	3	-	-	1	7
TUL-629	-	-	9	-	1	10
TUL-879	-	2	-	-	-	-
Totals		3	6	9	2	23

KEY: Bead types per Bennyhoff and Hughes 1987; A1a — Small spire-lopped, E2b — Thick-lipped, G1 — Tiny saucer, H2 — Disks drilled w/ metal needles, and K1 — Callus cup beads.

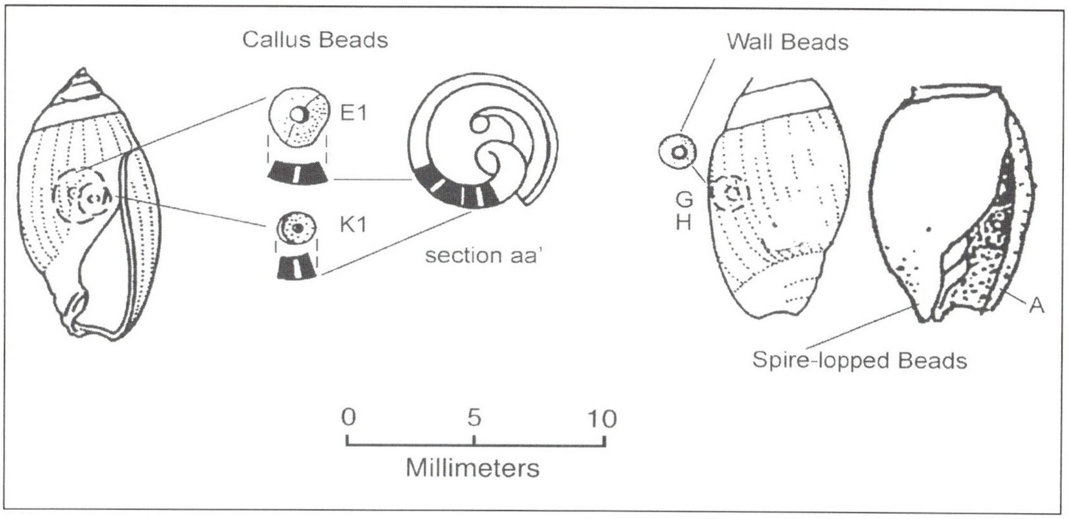

Figure 4.1 *Olivella* Bead Types: Style, Forms, and Manufacturing Methods.

Stone Disk Beads

Forty-six stone disk beads were recovered from 11 study sites. These beads are manufactured from rocks and minerals identified variously as talc/serpentinite (steatite) and perhaps dolomite. The latter are presumed to be the light, offwhite or cream-colored beads, and the former exhibit a wide variety of colors ranging from black through subtle shades of blue, green, and gray. These beads are mostly small (4–7 mm in diameter), monoconically drilled disks of good-quality stone of uniform texture.

All but two of the stone beads came from study sites located west of the Sierra crest. Crestal sites are virtually devoid of these beads. A review of information about eastern California bead distribution (Milliken 1999) indicates an almost complete absence of stone beads from the far southernmost portions of the Owens Valley. For Rose Valley and the Coso Volcanic Field areas, only five such beads have been discovered, and these were recovered from the Rose Spring site (INY-372) (Lanning 1963).

Radiocarbon dates from the TUL-879 site on the Kern Plateau provide an estimate for the initial date of these beads at ca. A.D. 1600 (Garfinkel et al. 1984), a date that is largely consistent with independent estimates indicating that these beads did not come into common use until the Late Period, Phase 2 in central California (Bennyhoff 1994:68; Moratto 1972:348, 1984:317). Furthermore, according to Gibson (1975), manufacture of these stone disk beads terminates about A.D. 1810. In the Hidden Reservoir area of Madera County, stone disks were rapidly replaced by glass trade beads (Fenenga 1977).

Researchers believe that stone disk beads came into the study area through trade with groups residing in the southern Sierra Nevada foothills (cf. Moratto 1988:202–217). Bead manufacture and use appear to have largely coincided with Foothill Yokuts territories where outcrops of good-quality steatite were located. Ethnographic data indicate that the Yokuts manufactured steatite disc beads, then used them as money and as grave offerings (Driver 1937:125; Gayton 1948:191; T. King 1968). Steatite disk beads are distributed mainly in the southern and central Sierra Nevada and in adjacent areas of the San Joaquin Valley (Moratto 1988).

Glass Trade Beads

Two hundred eighty two glass beads were found at 12 sites in the study area (Tables 4.1 and 4.3). Glass beads occur in many shapes, sizes, and colors throughout California and the Great Basin. Historic glass trade beads are hallmarks of the terminal Chimney Period. Many complex typological systems have been offered for the classification of glass beads (Bass and Andrews 1977; Gibson 1975; Kidd and Kidd 1970; Meighan 1955; Ross 1990; Titchenal 1994). Perhaps the most relevant study is that conducted by Titchenal (1994). Titchenal seriated 157 glass beads collected from 21 sites in the nearby Owens Valley. Titchenal's studies suggested six successive, time-sensitive, 19th-century, bead assemblages comprising various forms. Milliken (1999), expanding on that research, incorporated additional data derived from the Alabama Gates study (Delacorte 1999).

Most of the glass beads from the study area were recovered from sites on or near the Sierra Nevada crest (Scodie Mountain and Lamont Meadow sites). One component of KER-1298, located near McIvers Spring in the Scodie Mountains, yielded the largest glass bead collection (234 beads). At least 50 of those beads were highly fragmented examples, so small and fire-burned that they could not be confidently classified into particular types. California and Great Basin peoples customarily made offerings of beads for the dead and those fragmentary beads could represent a mourning ritual or cremation (Steward 1938; Zigmond 1968). The remaining beads from that site appear most consistent with Titchenal's Complex C dating from A.D. 1849 to 1856. Most of those beads (124) were small, light-blue, hot-tumbled forms. The second most common bead form was the medium-sized, translucent, cobalt blue, faceted and nonfaceted hexagonal form, Titchenal's CMS7b1 and CMG7b1 types, comprising some 48 items. These beads date most commonly from A.D. 1856 until 1864, but also occur with less frequency from A.D. 1849 until 1856.

Phoenix Button

A single button of brass with the words "JE RENAIS DE MES CENDRES" ("I rise from my ashes") was recovered from site KER-1286 in the Scodie Mountains. A mythological phoenix is depicted along with "NO. 9." Phoenix buttons have been described by E. Strong (1960, 1975) and Carrico (1982).

These buttons were originally made in London, England, and were manufactured for King Christophe, the ruler of Haiti from 1811 until 1820 (Carrico 1982; Dietz 1976; E. Strong 1975). The buttons were intended for military uniforms, the number 9 representing a specific military regiment, the Port de Paix. King Christophe died prematurely by self-inflicted wounds in 1820. The buttons sat in warehouses some years afterward and then were sold to American traders (E. Strong 1960, 1975). Subsequently, the buttons were used by merchants trading with the Indians throughout the Northwest from 1832–1835. These buttons were also important trade objects during the Mexican period in California history (Carrico 1982; E. Strong 1975). Phoenix buttons serve as valuable time markers because they were used for only a brief period and the button found in the Scodie Mountains probably dates between 1830 and 1835.

Dating of Site Components

Based on the various series of hydration readings, their associated coefficient of variation values, and data from other temporal indicators, 41 sites have single components, 14 have dual components, 20 are multiperiod loci, and 24 are indeterminate. Further, the ages of the various loci indicate that during the Kennedy Period, there are no single-component loci and only one dual-component locus. That single two-period locus is located in Kennedy Meadows.

In the subsequent Lamont Period, the number of single- or dual-period loci increases to three. All are located in Kennedy Meadows or in the Rockhouse Basin. In the following Canebrake Period, 10 single- or dual-period loci were found in a wide variety of settings in the areas of Bear Mountain, Rockhouse Basin, and the Scodie Mountains.

The Sawtooth Period witnessed a significant spike in the number of single- and dual-period loci. Twenty-six loci were situated throughout the study area. The Chimney Period is represented by 25 loci most of which (15) are located along the Sierra Crest in the Scodie Mountains and Morris Peak areas. Multiple-period, temporally mixed assemblages are largely restricted to areas outside of the Sierra Crest. Several such loci show occupations spanning the entire spectrum of prehistoric occupation of the Kern Plateau from the Terminal Pleistocene era in the Kennedy Period throughout the Holocene era to the most recent Chimney interval (e.g., TUL-488N, TUL-629 and TUL-621).

The most striking observation emerging from the review of the study site chronological data is the overabundance of Chimney Period artifacts. The magnitude of this intensification may be amplified since older components were probably obscured by younger ones. Additionally, scavenging/reworking of artifacts and differential preservation tend to skew the relative changes toward increased numbers of younger artifacts and settlements. Nevertheless, more than half (119) of the entire projectile point inventory (222) dates to this interval.

Considering that the Chimney Period represents only a small portion (550 years) of the total time range of aboriginal occupation within the study area (13,500 years), this differential representation of recent dating artifacts is all the more striking. In many piñon zones in the western Great Basin, similar findings have been reported where increased frequencies of post-650 B.P. projectile points reflect a pattern of continuing intensification in the use of upland resources — particularly piñon nut exploitation (Bettinger 1975; Delacorte 1990). This trend is suggested as having culminated in the pattern observed ethnographically, with piñon nuts representing a major food staple and critical component in the subsistence-settlement pattern (Steward 1933, 1938). Such a pattern strongly suggests increasing pressures from expanding Native populations to exploit more marginal and labor intensive resources.

Radiocarbon data provide further support for this pattern, with most dates (24 of the 28) falling within the latest period of occupation. Glass and stone beads (279 and 46, respectively) also fall into this interval and, as expected, are much more plentiful than the number of shell beads (24). Obsidian hydration readings largely conform to this pattern as well with 22 readings per 100-year interval for the Chimney Period and a nearly equivalent rate of 23.4 rims per century associated with the prior Sawtooth Period. Earlier periods have far fewer hydration measures represented, with only 6.6 values per hundred years in the Canebrake Period and less than one reading per century for the Lamont and Kennedy Periods.

Summary

A variety of chronological data provide sufficient information to date most of the sites in the study area. Obsidian hydration plays an important role. The 475 obsidian hydration readings greatly aid in site age determination. Development of a Coso obsidian Kern Plateau hydration rate provides an approximate chronological placement for the majority of the cultural deposits. The rate was found to work fairly well in placing the sites in a reasonable time sequence and in correctly attributing diagnostic projectile points to the most widely accepted temporal intervals.

Analysis of the chronological data from 77 sites revealed 56 single or dual period loci, as well as 19 mixed deposits and 23 site loci that were not dateable. These data allow 56 loci to be placed into one or two of the five chronological periods identified for the study area. These data also confirm that the study area was occupied from the end of the Pleistocene era until the historic contact period. The bulk of the occupations occurred during the two most recent chronological intervals: the Sawtooth (A.D. 600–1300) and Chimney (A.D. 1300–post contact) periods. The greatest number of time-sensitive artifacts and radiocarbon dates also occur in the Chimney Period.

Of 28 radiocarbon dates, 24 were from the Chimney Period and four date to the Sawtooth Period. Nearly 400 sherds of Owens Valley Brownware were retrieved and they also mark the Chimney Period and date from ca. A.D. 1500 to the historic era. Of the 351 glass, stone, and shell beads, most date to the closing centuries of the Chimney Period.

Projectile point types run the gamut and represent most of the time-marker forms typically found in the Great Basin. Fifty-one sites contained 222 classifiable points. Of these, 116 were members of the Desert series; 45 were of Rose Spring or Eastgate types; 27 fit the Humboldt series; 19 were specimens of the Elko series; two were Gypsum forms; nine are Pinto or Little Lake points, and two have been classified as Paleoindian Concave–based points.

A similar pattern of late prehistoric land use is recognized in the piñon zones of many western Great Basin areas. This widespread trend culminated in the pattern of ethnographic piñon use, with these nuts regarded as one of the most important foodstuffs and a dietary staple for Great Basin native peoples.

Chapter 5
Prehistoric Settlement Types, Territory, and Boundaries

Scope and Purpose

In this chapter, systematic analysis of the archaeological deposits and cultural materials allows us to suggest descriptive classes of settlement. The structure of the material remains aids in the definition of the prehistoric activities represented at the sites. A brief review of previous settlement taxonomies opens the discussion. Following that, a comprehensive classification, using qualitative characteristics, is presented. Next, quantitative measures are employed to refine these initial groupings. Discussion then turns to issues of general land-use patterns and prehistoric territoriality. Finally, the ethnic/linguistic attribution of particular groups of sites is considered. The spatial distribution and dating of archaeological components provide evidence of the timing and character of prehistoric population movements and *in-situ* developments in the study area. The information presented in this chapter lays the groundwork for evaluation of linguistic prehistory to be taken up in Chapter 6.

Classification of Site Loci

All seven of the prior studies in the research area have classified archaeological deposits. The more comprehensive of these studies have categorized most of the loci examined in this book. Six of the studies have used qualitative surface characteristics. One study also used a computer to process "objective" data (McGuire and Garfinkel 1980). Rigorous quantitative and statistical analyses are inappropriate for the subject components because their surface artifacts were only partially collected and in many cases were not completely identified.

Sampling strategies differed among researchers, with some sites having their surface cultural remains completely documented (Garfinkel et al. 1980; McGuire 1983) while others were only partially sampled (Ambro et al. 1981; Bard et al. 1985; Garfinkel et al. 1984; McGuire 1981; McGuire and Garfinkel 1980). Therefore, the frequencies of artifacts are not directly comparable. An exception was the documentation of rock rings, bedrock milling features, and portable milling equipment, which were fully tallied and described by all of the previous investigators.

Prehistoric sites are often an amalgam of various aboriginal occupations differing in age and function. Prehistorians working in eastern California often have recognized that significant overprinting of different settlement types occurs at many archaeological sites and may compromise reconstructions of settlement-subsistence change (cf. Basgall and Giambastiani 1995; Hildebrandt and Ruby 1999). I attempt to differentiate various areas of the sites into distinct loci of relatively uniform character and assumed function in order to provide valid inferences regarding their age and function.

A descriptive typology of site components (site loci) was selected as the most useful classification system. Additionally, excavation data further refined the descriptive typology. As with prior treatments, the present analysis divides loci into four mutually exclusive classes (Bard et al. 1985; Garfinkel et al. 1984). The categories used in this analysis are similar to those usually employed to classify aboriginal settlements in the upland piñon forests of eastern California (Bettinger 1975, 1979, 1982; Delacorte

1990; Hildebrandt and Ruby 1999). Site classifications are typically based on both their archaeological assemblages and natural settings. Resultant taxonomies are similar to settlement forms described ethnographically (Steward 1938; Voegelin 1938).

Variables selected for the first-order groupings appear to have important functional implications and could be used as relevant sorting criteria. Three characteristics met those criteria. They are presence/absence of (1) rock rings, (2) bedrock milling features or portable milling equipment, and (3) midden. These attributes differentiate loci into classes that mimic intuitive and computer-generated site types.

Rock Rings

Rock rings are highly correlated with piñon procurement and in many instances are the remains of piñon caches located within upland piñon-juniper forests. The consensus is that such caches are the remains of facilities used for the storage of green piñon pine cones containing nuts (Bettinger 1989; Hildebrandt and Ruby 1999; Zeanah 2002). Ruhstaller (1980), summarizing existing information on study area features of this type, recognized that the interior diameters of most rock rings averaged two meters or less and had little in the way of associated cultural materials. These circular configurations of stone rest upon rocky, granitic substrate (most likely to discourage disturbance from below by rodents) and often have large concentrations of unmodified quartz cobbles in association. These rock ring features probably are the remains of piñon caches that, after being opened, leave the circular rings of stones with their interior contents removed.

Larger stone rings, averaging more than 2 m in diameter, sometimes with gaps or "doorways," are found less frequently. Recovered within these more sizeable rings are artifacts associated with domestic activity (spent stone tools, projectile points, beads, pottery sherds, etc.). These rings probably served as the bases or structural supports of temporary houses or other shelters of brush.

Milling Implements

Milling tools, including hand stones, milling slabs, portable or bedrock mortars/metates, and pestles, are commonly associated with the milling of various nuts and hard seeds. The presence of such artifacts and features is presumed to indicate reliance on vegetal foods and their processing.

Millingstones and hand stones are common at the Kern Plateau piñon zone study sites. Such stones are also ubiquitous in the upland piñon forests of the Owens Valley, Deep Springs Valley, and Coso Range of eastern California (Bettinger 1975; Delacorte 1990; Gilreath and Hildebrandt 1997). In many other areas of the Great Basin, that is not the case (McGuire and Garfinkel 1976; Thomas 1971; Zeanah 2002). One explanation for the difference is that in those other areas piñon nuts were normally transported, unshelled in their cones, back to base camps. That strategy dismissed the need for milling equipment to hull the nuts from the cones at the locations of piñon nut procurement (Zeanah 2002).

Such logistical rather than residential use of the upland piñon forests may have been a function of the relative productivity of piñon zones in those areas. Piñon stands in the southern, central, and western Great Basin appear to be denser and more prolific than in other areas of the Great Basin (Zeanah 2002). Richness of piñon has been shown to correlate with the relative density of groundstone tools across the Great Basin and in the far southern Sierra Nevada (Zeanah 2002).

Midden

The presence and depth of anthropic soils (midden deposits) are usually good indicators of the character and intensity of a particular site's occupation. Repetitive use of particular locations, especially when organic refuse is left to decompose, results in midden accumulation. A midden deposit can be one indicator of a lengthy occupation. Midden sites containing milling equipment probably would have been places reoccupied during years when piñon crops were especially fruitful. Such occupation sites were often developed to allow groups to overwinter in sheltered locations near a permanent source of water, relatively free from snow, during exceptionally rich piñon years. Such a strategy is inherently more efficient than attempting to move surplus nuts to more distant villages (Garfinkel et al.1984: 13–20).

These three characteristics serve as differentiating criteria for the four classes of archaeological loci. Initially, deposits may be divided into those with rock rings and those without. Deposits lacking such facilities are then classified as to whether they evince milling activity. Lastly, a division is made between loci with and without midden. Using such divisions, site loci can be segregated into four groups:

Class 1: Rock ring loci
Class 2: Midden deposits with milling implements
Class 3: Loci with milling implements but lacking midden
Class 4: Loci with cultural material, principally debitage (chipping waste from flaked-stone tool production) and flaked-stone implements (projectile points, formalized tools and utilized flakes) but lacking midden and milling implements

Class 1 Loci: Class 1 loci contain rock rings as their principal distinguishing criteria. Twenty-one (22%) of the 96 loci contain such features (Table 5.1). Most sites contain no more than a single rock ring, yet one (TUL-484) has six such features. Altogether, 29 circular rock features were found. The majority, 25 (86%), are in the Morris Peak/Lamont Meadow and Scodie Mountains segments of the Pacific Crest Trail, on the crest of the Sierra Nevada (Ambro et al. 1981; Garfinkel et al. 1980; McGuire 1983).

Class 2 Loci: Twenty-seven loci (28%) exhibit milling implements and associated midden deposits (Table 5.1). Midden depths varied from 0.2 to 1.35 meters. Twenty-two loci (81%) have bedrock milling features. Ten (37%) also produced sherds of Owens Valley Brownware.

Class 3 Loci: Twenty-two loci (23%) contain milling features and artifacts, yet lack midden deposits characteristic of the more intensive occupations of Class 2 loci (Table 5.1). Deposits still, in some cases, contained buried cultural materials to a depth as great as 0.85 meters. Ceramic sherds are much less plentiful, with only five of the 22 deposits (23%) having yielded associated pottery.

Class 4 Loci: Twenty-four (25%) deposits are classified as Class 4 loci (Table 5.1). None of these loci have yielded ceramics or beads. Class 4 deposits are often only a collection of flaked-stone debris. They range from quite small (20 m^2), diffuse (0.3 artifacts/m^2) "flake scatters" to large (230,000 m^2) and more dense arrays (up to 1000 flakes/m^2 in some areas) of toolstone debitage that are sometimes described as "flaked-stone workshops."

Dating the Loci

Forty single-period and 11 dual-period loci are represented. Twenty-one loci have multiperiod deposits and the balance (24) are of indeterminate age. Several diachronic trends regarding the age of these loci

Table 5.1 Surface Characteristics of Kern Plateau Loci: Scodie Mountains.

SIZE	ELEV.		RK	MILL.	MILL.								DEPTH	LOCUS	
SQ. M.	M. AMSL	SITE NO.	RNGS	BED.	POR.	CER.	PTS	UNI.	BIF.	DEB.	BDS	CRNS	MIDD	(CM.)	TYPE	PERIOD
2500	1980	KER-1269	0	17	2	0	1	0	0	4	0	0	-	0	3	Saw./Cnbrk
650	1995	KER-1270	0	0	0	0	0	0	0	71	0	0	-	0	4	Cnbrk
3200	1625	KER-1971	0	0	0	0	0	0	0	14	0	0	-	0	4	Cnbrk
8000	2075	KER-1296	0	0	0	0	2	0	0	128	0	0	-	50	4	Saw.
5600	2075	KER-1297	0	8	0	75	3	1	7	230	1	0	-	25	3	Multiple
5600	2075	KER-1298	0	12	3	29	7	1	2	998	2	0	+	40	2	Multiple
2700	2075	KER-1299D	0	2	0	0	3	0	0	104	1	0	+	40	2	Saw.
3200	2075	KER-1299E	0	0	0	0	0	0	0	76	0	0	+	20	4	Cnbrk
3000	2075	KER-1299F	0	14	2	1	1	0	2	106	0	0	+	30	2	Ch./Saw.
15	2075	KER-1299G	2	0	0	0	0	0	0	2	0	0	-	30	1	Insuff.
1000	2040	KER-1272	0	0	1	0	0	0	0	39	0	0	-	10	3	Insuff.
600	1975	KER-1273	0	12	5	0	0	0	0	29	0	0	-	40	3	Chimney
600	2090	KER-1274	1	0	0	0	0	0	0	30	0	0	-	0	1	Insuff.
7200	2095	KER-1275	0	0	2	62	1	0	0	26	0	1	-	45	3	Ch./Saw.
2000	2065	KER-1276A	0	1	2	0	0	0	1	34	0	0	-	20	3	Chimney
100	2065	KER-1276B	1	0	0	0	1	0	0	7	0	0	-	30	1	Chimney
3400	2120	KER-1277	0	0	0	0	1	0	3	276	0	0	-	75	4	Multiple
100	2100	KER-1278	0	0	0	0	1	0	0	4	0	0	-	0	4	Ch./Saw.
1400	1940	KER-1279	0	3	2	0	0	0	0	30	0	0	+	70	2	Cnbrk
1100	1950	KER-1280	0	2	1	0	0	0	0	15	0	0	-	20	3	Cnbrk
28000	2135	KER-1281	1	32	0	0	7	2	1	633	0	0	+	25	1	Multiple
5400	2060	KER-1282	0	0	6	0	0	0	1	60	0	0	-	30	3	Cnbrk
1600	1920	KER-1283	0	5	11	2	9	0	0	98	1	0	+	40	2	Multiple
2100	2100	KER-1284	0	1	1	0	1	1	0	22	0	0	-	30	3	Saw.
1200	2100	KER-1285	1	1	0	0	0	0	0	22	0	0	-	30	1	Saw.
8000	1975	KER-1286	3	0	2	12	0	0	0	3	1	0	-	30	1	Ch./Saw.

Table 5.1 Surface Characteristics of Kern Plateau Loci (continued): Morris Peak and Lamont Meadow.

SIZE	ELEV.		RK	MIL.	MILL.								DEPTH	LOCUS	
SQ. M.	M. AMSL	SITE NO.	RNGS	BED	PORT	CER.	PTS	UNI.	BIF.	DEB.	BDS	CRNS	MIDD	(CM.)	TYPE	PERIOD
50	1890	KER-748	1	0	2	0	1	0	1	28	0	0	+	10	1	Chimney
800	2010	KER-744	1	0	3	0	2	0	0	65	0	0	+	20	1	Insuff.
1500	2035	KER-743	2	0	5	0	1	0	4	79	0	0	+	30	1	Chimney
10000	1985	KER-742	1	0	0	1	1	0	10	502	0	0	+	30	1	Chimney
13	1980	KER-741	0	0	0	0	0	0	0	0	0	1	-	0	4	Insuff.
25	2210	KER-747	1	0	0	0	0	0	0	0	0	0	-	0	1	Insuff.
13	2165	KER-746	1	0	0	0	0	0	0	0	0	1	-	0	1	Insuff.
13	2165	KER-738	1	0	0	0	0	0	0	0	0	0	-	0	1	Insuff.
15	2075	KER-745	1	0	0	0	0	0	1	0	0	0	-	0	1	Insuff.
300	2165	KER-737	1	0	1	0	3	0	7	364	0	0	-	0	1	Ch./Saw.
6500	2100	TUL-484A	5	0	1	0	2	0	10	60	0	0	-	0	1	Chimney
12000	2100	TUL-484B	1	0	1	0	4	0	2	422	0	0	+	20	1	Saw.
21500	2100	TUL-484C	0	0	0	0	4	0	1	224	0	0	+	20	4	Saw.
100	2100	TUL-484D	0	0	0	0	0	0	10	152	0	0	-	?	4	Insuff.
8000	1930	TUL-483	0	0	7	3	14	4	0	804	0	0	-	20	2	Ch./Saw.
1200	1770	TUL-482	0	0	1	5	2	2	4	301	0	0	-	30	3	Chimney
2700	1770	TUL-481	0	0	8	9	6	0	3	58	2	0	+	30	2	Ch./Saw.
100	1650	TUL-480	0	0	0	0	0	0	0	32	0	0	-	0	4	Insuff.
10500	1650	TUL-487	0	0	0	0	0	0	1	88	0	0	+	40	4	Insuff.
9100	1650	TUL-485	0	19	3	0	1	0	1	438	0	0	-	40	3	Chimney
25000	1740	TUL-488N	0	4	2	0	7	1	10	938	0	0	+	135	2	Multiple
100	1760	TUL-489	0	5	1	0	0	0	1	18	0	0	+	30	2	Insuff.

Table 5.1 Surface Characteristics of Kern Plateau Loci (continued): Bear Mountain.

SIZE	ELEV.		RK	MIL.	MILL.									DEPTH	LOCUS	
SQ. M.	M. AMSL	SITE NO.	RNGS	BED.	PORT.	CER.	PTS	UNI.	BIF.	DEB.	BDS	CRNS	MIDD.	(CM.)	TYPE	PERIOD
6000	1980	TUL-620	0	2	10	0	8	1	7	3694	7	0	-	0	3	Ch./Saw.
1400	2000	TUL-621	0	51	0	4	5	0	3	2072	7	0	+	50	2	Multiple
2500	2000	TUL-629	0	1	10	0	3	1	9	787	4	0	+	60	2	Multiple
3600	2100	TUL-617	0	0	0	0	4	0	7	596	1	0	-	0	4	Multiple
2000	2100	TUL-623	0	0	3	0	4	0	2	553	0	0	+	50	2	Chimney
7800	2200	TUL-630	0	5	0	0	1	0	8	697	0	0	+	80	2	Canebrake
200	2230	TUL-625A	0	5	2	0	3	0	1	72	3	0	+	20	3	Chimney
50	2230	TUL-625B	1	0	0	0	0	0	0	132	0	0	-	30	1	Sawtooth
2000	2235	TUL-618	1	2	1	0	1	0	7	1322	2	0	-	50	1	Insuff.
500	2415	TUL-616	0	0	0	0	1	0	3	72	0	0	+	20	4	Lamont
2000	2430	TUL-628	0	0	0	0	1	0	1	1322	0	0	-	0	4	Insuff.
3000	2375	TUL-767	0	0	0	0	1	0	2	584	0	0	-	0	4	Sawtooth
13,000	2200	TUL-636	0	23	1	0	5	0	11	5269	0	0	+	20	2	Multiple
800	2100	TUL-634	0	0	0	0	1	0	3	87	0	0	+	30	4	Ch./Sawtooth
100	1980	TUL-632	1	0	0	0	1	0	2	4	1	0	-	0	1	Insuff.
3600	2250	TUL-619	1	0	0	0	3	0	4	270	0	0	+	80	1	Sawtooth

Table 5.1 Surface Characteristics of Kern Plateau Loci (continued): Rockhouse Basin/Kennedy Meadows.

SIZE	ELEV.		RK	MIL.	MILL.									DEPTH	LOCUS	
SQ. M.	M. AMSL	SITE NO.	RNGS	BED.	PORT.	CER.	PTS	UNI.	BIF.	DEB.	BDS	CRNS	MIDD.	(CM.)	TYPE	PERIOD
17,000	1780	TUL-877	0	16	0	0	3	0	4	256	0	0	+	60	2	Multiple
50,000	1730	TUL-878	0	17	0	0	0	0		P	0	0	-	0	3	Insuff.
10,000	1830	TUL-879A	0	5	0	1	4	1	5	168	0	0	+	110	2	Multiple
100	1830	TUL-879B	0	0	1	8	0	0	3	36	0	0	+	?	2	Chimney
230,000	1730	TUL-880	0	0	0	0	0	0	0	P	0	0	-	0	4	Insuff.
200	1730	TUL-895	0	13	0	0	1	1	1	P	0	0	+	?	2	Cnbrk
190,000	1710	TUL-511	0	66	1	70	3	1	1	P	2	0	+	?	2	Multiple
10,000	1730	TUL-888	0	0	0	0	0	1	1	P	0	0	-	0	4	Insuff.
80,000	1920	TUL-882	0	10	5	0	3	1	1	P	0	0	-	0	3	Multiple
10,000	1900	TUL-881	0	0	0	0	0	1	1	P	0	0	-	0	4	Insuff.
2500	1800	TUL-883	0	0	0	0	0	1	1	P	0	0	-	0	4	Insuff.
3000	1775	TUL-884	0	2	0	0	1	1	1	P	0	0	-	85	3	Multiple
8000	1805	TUL-887	0	4	0	0	1	0	0	17	0	0	+	40	2	Sawtooth
4500	1775	TUL-885	0	1	0	0	0	0	1	P	0	0	-	0	3	Insuff.
1300	1800	TUL-886	0	0	0	0	0	0	0	P	0	0	-	0	4	Insuff.
800	1830	TUL-890A	0	0	0	33	0	0	0	235	0	0	+	50	4	Ch/Swtth
3700	1830	TUL-890B	0	7	0	1	2	1	6	1745	0	0	+	80	2	Sawtooth
1000	1830	TUL-890C	0	0	0	0	0	0	1	167	0	0	+	30	2	Sawtooth
3150	1830	TUL-889A	0	6	0	0	1	0	19	3687	0	0	-	70	3	Cnbrk/Lmnt

Table 5.1 Surface Characteristics of Kern Plateau Loci (continued): Rockhouse Basin/Kennedy Meadows.

SIZE	ELEV.		RK	MIL.	MILL.									DEPTH	LOCUS	
SQ. M.	M. AMSL	SITE NO.	RNGS	BED.	PORT.	CER.	PTS	UNI.	BIF.	DEB.	BDS	CRNS	MIDD	(CM.)	TYPE	PERIOD
2225	1830	TUL-889B	0	0	0	0	1	0	2	3319	0	0	-	30	4	Cnbrk
5000	1830	TUL-889C	0	0	0	0	2	0	3	1466	0	0	-	50	4	Cnbrk/Lmnt
20	1830	TUL-889D	0	6	5	0	0	0		1	0	0	-	30	3	Insuff.
4000	1830	TUL-891	0	0	3	11	0	0	2	1162	0	0	-	0	3	Chimney
		Surface														
850	1830	TUL-891	0	34	0	0	0	0	0	0	0	0	+	65	2	Sawtooth
		Middens	0	0	0	0	0	0	0	0	0	0	+	65	2	Sawtooth
20	1830	TUL-891	0	0	0	0	0	0	0	P	0	0	+	40	3	Sawtooth
		Depression														
160,000	1830	TUL-894	0	4	2	0	1	1	1	P	0	0	+	35	2	Chimney
200	1828	TUL-895	0	13	0	1	0	0	0	P	0	0	+	?	2	Cnbrk
36,625	1800	TUL-896	0	0	0	0	7	0	0	7200	0	0	-	90	4	Multiple
49,225	1800	TUL-897	0	3	2	0	2	0	0	574	0	0	-	40	3	Lmnt/Knndy
52,000	1800	TUL-898	0	4	0	0	0	1	1	3144	0	0	+	70	2	Multiple
65,400	1800	TUL-899	0	0	1	0	11	3	6	23,678+	0	0	-	55	3	Multiple
7370	1800	TUL-909	0	9	0	0	1	3	1	180	0	0	+	40	2	Chimney

KEY: Size in square meters, elevation in meters (above Mean Sea Level), rock rings, bedrock milling, portable milling, ceramics, points, uniface, biface, debitage, beads, cairns, midden, centimeters.
Loci type: 1=rock rings, 2=midden, 3=milling and no midden, 4=lithic scatter
Periods: Chimney, Sawtooth, Canebrake, Lamont, Kennedy, insufficient data
P=present, A=absent

Table 5.2 Site Loci by Period and Type (Total Inventory)

	Age (years BP)	Classification				
		1 RR	2 MM	3 MNM	4 LS	Subtotal
Component Loci						
Chimney	<650	7	7	7	2	23
Sawtooth	650–1350	7	9	3	5	24
Canebrake	1350–3500	0	3	4	5	12
Lamont	3500–8500	0	0	2	1	3
Kennedy	>8500	0	0	1	0	1
Subtotal		*14*	*19*	*17*	*13*	*63*
Multiple Period		1	10	6	4	21
Insufficient Data		9	1	3	11	24
Total		*24*	*30*	*26*	*28*	*108*
Single Period				Loci Percentages		
Chimney	<650	30	30	30	9	100
Sawtooth	650–1350	29	37	12	21	100
Canebrake	1350–3500	0	25	33	42	100
Lamont	3500-8500	0	0	75	25	100
Kennedy	>8500	0	0	100	0	100

KEY: Class 1-RR — Loci containing rock rings; Class 2-MM — Loci with milling (bedrock or portable) and midden; Class 3-MNM — Components with milling lacking midden; Class 4-LS — Flaked stone scatters.

can be noted. In general, loci are predominantly Chimney and Sawtooth in age, consistent with the obsidian hydration data reported in Chapter 4. Yet, proportions of the types of sites found in these consecutive periods differ (Table 5.2).

All Class 1 rock ring loci date to either the Chimney or Sawtooth periods. No rock rings date to earlier periods. Also, cultural materials recovered from these deposits nearly always point to a single time period. Such loci appear to have been used only for brief, single episodes in time and apparently were rarely revisited or reused. In contrast, flaked-stone scatters and lithic workshops (Class 4) dominate the earlier Sawtooth and Canebrake periods. Such loci are largely absent in the following Chimney era. Residential components, Class 2 loci, are the most likely types of deposits to represent multiple-period expressions. Local environmental factors (a permanent source of water and naturally sheltered area) appear to have determined the highly favored localities that were occupied repeatedly throughout prehistory.

Subsurface Constituents

Many sites (54 of the 96 loci identified) yielded further information from subsurface study. Excavation data in most cases served to support the preliminary classification. Quantification of excavation data permits comparison within and between the various loci classes. The data also allow us to refine observations regarding their character and permit subdivision of these broad loci classes. Excavation data allow direct comparison between loci based on the frequency of particular types of cultural material as expressed by the number of items retrieved per cubic meter of excavated deposit (Tables 5.3–5.6).

It has been recognized that toolstone manufacturing technology within Eastern California and the southwestern Great Basin can be characterized by a distinct set of changes. An initial emphasis (ca. 8000–1000 B.P.) on biface preform or flake-blank technology eventually gave way to flake-based reduction techniques during late prehistoric times. Earlier-period flaked-stone assemblages include large

Table 5.3 Excavation Data: Class 1 (Rock Ring Loci)

(Total Items and items per cubic meter of deposit)

Loci*	Total Volume of Excavated Deposit	Debitage		Stone Tools		Milling Tools		Faunal Remains		Beads		Pottery	
		n	n/m3	n	n/m3	n	n/m3	n	n/m3	n	n/m3	n	n/m3
K-1276	0.4	7	17.5	0	0	0	0	0	0	0	0	0	0
K-1281	1	423	423	21	21	0	0	0	0	0	0	0	0
K-1286	1.5	69	46	8	5.3	1	0.7	14	9.3	3	2	7	4.6
K-748	0.4	12	30	4	10	0	0	3	7.5	1	2.5	0	0
K-744	0.5	68	136	0	0	0	0	0	0	0	0	0	0
K-743	2.4	326	135	10	4.2	7	2.9	18	7.5	0	0	0	0
K-742	1.1	487	443	53	48.2	1	0.9	6	7	0	0	1	0.9
K-737	0.1	9	90	19	190	1	0.1	0	0	0	0	0	0
T-619	0.8	173	216	2	2.5	0	0	0	0	0	0	0	0
T-618	1	300	300	1	1	0	0	0	0	0	0	0	0
T-625	1.1	135	123	0.9	0.9	0	0	0	0	0	0	0	0
Range		17.5–443	0–190		0–2.9		0–9.3		0–2.5		0–4.6		
Mean		178.1	31.4		1.15		7.8		0.3		2.7		

KEY: n=total items, n/m3=totals per cubic meter of deposit, a measure that indicates the mean number of artifacts or animal bones retrieved per cubic meter of sampled deposit. The measure is calculated on the available information for a specific locality. Range and mean apply only to the volumetric figures.
*K=Kern and T=Tulare

bifaces that decrease in abundance, size and formality and are ultimately replaced by more numerous flake-based tools.

The early toolstone-reduction technology may be found in the voluminous debitage and ubiquitous biface thinning debris that characterize older settlements. After the introduction of the bow and arrow (ca. A.D. 600) and especially with the advent of smaller arrow points of the Desert Series (beginning at A.D. 1300), a dramatic drop in the quantity and size of flaked-stone debris is recognized (Delacorte 1999; Delacorte et al. 1995; Gilreath and Hildebrandt 1997; E. Skinner 1986). Hence the lack of flaked-stone debris is predictable given the age and character of Class 1 rock ring loci. This shift in lithic reduction strategies is thought to relate to more than just simple technological considerations and appears to be a function of several factors including settlement centralization, minimization in foraging range, and reductions in residential mobility (Delacorte 1999; Garfinkel et al. 2004; Gilreath and Hildebrandt 1997).

Class 2 Loci: Class 2 loci provide the greatest amounts of cultural materials, both with respect to the various artifact categories and with reference to vertebrate faunal remains. Yet these loci are also quite varied, suggesting that they differed in several ways. Prior studies have in fact suggested a separation,

Table 5.4 Excavation Data: Class 2 (Milling Equipment With Midden).

(Total Items and items per cubic meter of deposit)

Loci*	Total Volume of Excavated Deposit	Debitage		Stone Tools		Milling Tools		Faunal Remains		Beads		Pottery	
		n	n/m3	n	n/m3	n	n/m3	n	n/m3	n	n/m3	n	n/m3
T-481	0.6	257	428	30	50	0	0	22	13	8	13.5	6	10
T-488N	12.2	19517	1616	314	26	24	2	1742	159	20	1.6	3	0.2
T-489	2.5	300	120	11	4.4	4	1.6	0	0	0	0	0	0
T-629	2.1	24261	11553	291	139	13	6.2	86	41	30	14.3	1	0.5
T-621	3.7	23865	6450	424	115	4	1.6	422	114	61	1.6	0	0
T-630	0.8	769	961	2	2.5	0	0	0	0	0	0	0	0
T-636	0.6	135	123	21	35	0	0	0	0	0	0	0	0
T-623	0.5	300	600	19	55	110	1	2	0	0	0	0	0
K-1299C	1.6	592	370	17	11	3	1.8	546	341	41	25.6	0	0
K-1299D	1.1	213	194	1	1	0	0	0	0	0	0	0	0
K-1299F	0.6	80	133	1	1.6	0	0	9	15	1	1.6	0	0
K-1279	0.7	114	163	2	2.8	0	0	0	0	0	0	0	0
K-1283	0.8	15	18.7	3	3.8	0	0	34	42.5	0	0	0	0
T-877	4.5	1350	300	11	2.4	5	1.1	2	0.4	0	0	0	0
T-879A	15.9	4193	264	119	7.5	3	0.2	2652	167	26	1.6	20	1.3
T-894	1.7	778	458	4	2.4	1	0.6	0	0	0	0	0	0
T-898	3.1	1550	500	11	3.5	10	3.2	90	39	2	0.6	1	0.3
T-909	1.3	625	481	1	0.8	0	0	0	0	1	0.76	0	0
Range		18.7-11553		.8-139		0–6.2		0–341		0–25.6		0–10	
Mean		1355		31		0.5		92.2		6.8		2.5	

KEY: n = total items, n/m3 = totals per cubic meter of deposit, a measure that indicates the mean number of artifacts or animal bones retrieved per cubic meter of sampled deposit. The measure is calculated on the available information for a specific locality. Range and mean apply only to the volumetric figures.
*K = Kern and T = Tulare

labeling deposits as either base camps or temporary camps (Garfinkel et al. 1980; McGuire and Garfinkel 1980).

Base camps themselves appear to vary. Some deposits contain only high densities of flake waste and others contain large quantities of both debitage and faunal material. Yet others have only high densities of faunal remains. Sites in the first subgroup include TUL-636 and TUL-623. The second group includes TUL-488, TUL-629, and TUL-621.

The third group encompasses KER-1298, KER-1283 and TUL-879A. All of these base- camp-like Class 2 deposits contain either more than 900 items of debitage per cubic meter of deposit *or* 40 or more animal bones per cubic meter of excavated midden.

Table 5.5 Excavation Data: Class 3 (Milling Equipment Without Midden).

(Total Items and items per cubic meter of deposit)

Loci***	Total Volume of Excavated Deposit	Debitage		Stone Tools		Milling Tools		Faunal Remains		Beads		Pottery	
		n	n/m3	n	n/m3	n	n/m3	n	n/m3	n	n/m3	n	n/m3
T-481	0.2	6	30	0	0	0	0	0	0	0	0	0	0
T-482	0.9	201	223	13	14	0	0	33	37	1	1.1	3	3.3
T-485	0.9	421	468	4	4.4	3	3.3	3	3.3	1	1.1	0	0
K-1269	0.3	9	30	3	10	0	0	0	0	0	0	0	0
K-1299B	0.5	150	300	0	0	0	0	0	0	0	0	0	0
K-1273	0.4	50	125	2	5	0	0	0	0	0	0	0	0
K-1275	0.3	25	83.3	1	3.3	0	0	1	3.3	0	0	0	0
K-1276A	1.0	5	5	1	1	0	0	0	0	0	0	0	0
K-1280	0.3	22	73.3	0	0	0	0	0	0	0	0	0	0
K-1282	0.6	51	85	0	0	0	0	0	0	0	0	0	0
K-1284	0.6	5	8.3	1	1.6	0	0	0	0	0	0	0	0
T-884	3.3	533	162	3	0.9	0	0	40	12.1	0	0	0	0
T-889A	2.8	27494	9819	22	7.9	0	0	11	3.9	1	0.4	0	0
T-897*	1.8	?	234	0	0	0	0	0	0	0	0	0	0
T-899*	3.6	?	3193	1	0.3	1	0.3	0	0	0	0	0	0
Range			5-300**		0-14.4		0-3.3		0-37		0-1.1		0-3.3
Mean			141		4.9		1.8		11.9		0.9		3.3

KEY: n = total items, n/m3 = totals per cubic meter of deposit: This measure indicates the mean number of artifacts or animal bones retrieved per cubic meter of sampled deposit. The measure is calculated on the available information for a specific locality. Range and mean apply only to the volumetric figures.

*Actual debitage volumes could not be precisely quantified because information was not available from the original study.

**Excludes TUL-889A and TUL-899 (see discussion) ***K = Kern and T = Tulare.

In comparison, temporary camps lack an abundance of either debitage or faunal materials yet still exhibit a wide range of cultural remains and a midden suggesting some level of residential use. The following localities fit the latter description and better fall into temporary camp designations: TUL-481, TUL-489, KER-1299D, KER-1299F, KER-1279, TUL-877, TUL-898 and TUL-909.

Class 3 Loci: Except for two unusual loci (TUL-889A and TUL-899), Class 3 deposits are uniform in their low frequency of cultural remains. Debitage is the only category of cultural material to be represented in any abundance at such localities. These loci, while exhibiting surface milling features or portable milling equipment, lack subsurface groundstone tools, thus appearing to demonstrate that Class 3 loci were infrequently used milling localities.

TUL-889A and TUL-899 are exceptions to these characterizations. Their yield of subsurface flaked-stone materials is extraordinary, especially given the fact that both localities bear no evidence of a wider

Table 5.6 Excavation Data: Class 4 (Flaked Stone Scatters).

(Total Items and items per cubic meter of deposit)

Loci**	Total Volume of Excavated Deposit	Debitage		Stone Tools		Milling Tools		Faunal Remains		Beads		Pottery	
		n	n/m3	n	n/m3	n	n/m3	n	n/m3	n	n/m3	n	n/m3
T-487	1.2	263	219	4	3.3	0	0	6	5	0	0	0	0
T-634	0.3	30	100	0	0	0	0	0	0	0	0	0	0
T-616	0.2	30	150	1	5	5	0	0	0	0	0	0	0
K-1270	0.3	18	60	0	0	0	0	0	0	0	0	0	0
K-1971	0.2	1	5	2	10	0	0	0	0	0	0	0	0
K-1296A	0.6	756	1260	5	8.3	0	0	0	0	0	0	0	0
K-1277	1.1	434	394	3	2.7	1	0.9	3	2.7	0	0	0	0
T-889B	0.9	21	23.3	0	0	1	1.1	0	0	0	0	0	0
T-889C	1.4	429	306.4	1	0.7	0	0	0	0	0	0	0	0
T-889D	1	77	81	1	1	1	1	25	25	0	0	0	0
T-896*	2.9	?	?	0	0	0	0	1	<.1	0	0	0	0
Range			5-1260		0-10		0-1.1		0-25		NA		NA
Mean			254		4.4		1		10.9		NA		NA

<u>KEY</u>: n = total items n/m3 = totals per cubic meter of deposit. This measure indicates the mean number of artifacts or animal bones per cubic meter of sampled deposit.
The measure is a simple average based on the available information. Range and mean apply only to volumetric figures. *Actual debitage volumes could not be precisely quantified because data was absent in the original monograph.
**K = Kern and T = Tulare

range of cultural activities, nor do they possess a developed midden. TUL-899 may contain such an abundance of flaked stone partly as a function of its occupational history. TUL-899 appears to have been consistently reoccupied over a lengthy interval and has temporally sensitive projectile points and obsidian hydration rims representing recurrent use over 10,000 years.

Alternatively, both loci (situated on the east banks of the South Fork of the Kern River at 6000 feet in elevation) could have served as favorable upland game (mainly deer) intercept/hunting localities. TUL-899 yielded one of the highest frequencies (n = 11) of surface projectile points for any site recorded in the study area and contained a rather large quantity (n = 9) of bifaces. TUL-889A, while not exhibiting such a large number of surface points, did yield seven subsurface dart points and a large number (n = 20) of fragmentary bifaces.

TUL-889A may have been misclassified since it owes its designation as a Class 3 locus to its association with bedrock milling features. Such stationary milling implements are thought by many to date much later in time than most of the remains produced from TUL-889A (Stevens 2003). If the locus had lacked those features, it would have been identified as a Class 4 locus. Yet even when compared to other Class 4 loci, TUL-889A produced a surprising number of flaked-stone items (averaging nearly 10,000 pieces of debitage per cubic meter of excavated deposit) — the highest yield of any Class 3 or Class 4 deposit yet recognized in the study area.

This extraordinary volume of flaked-stone debris leads to the conclusion that TUL-889A might have been a major flaked-stone workshop. Previously it had been suggested that this site might have served as a "staging area" for the production of obsidian bifaces intended for long-distance, trans-Sierran exchange (Bard et al. 1985:13–20). It seems reasonable that TUL-889A could be associated with hunter/trader/travelers who favored certain locations on treks across the Sierra. Coso obsidian bifaces were common items of trade and TUL-889A functioned (at least in part) as a lithic processing center where bifaces were produced for immediate use and for export.

TUL-889A dates solely to the Canebrake and Lamont periods, and TUL-899 has 73% (27 of 37) of its hydration rims and all its projectile points dating from the Canebrake Period or earlier. Such large quantities of debitage would be a predictable consequence of the intensive reduction activities typical of the Canebrake Period when great numbers of large Coso obsidian bifaces were being produced for exchange at the nearby Coso obsidian quarries (Gilreath and Hildebrandt 1997). Coso obsidian working during the Canebrake Period was 10 times more intensive than during earlier or later periods (Hildebrandt and McGuire 2003: Figure 6).

Class 4 Loci: Class 4 loci are composed almost exclusively of flaked-stone materials. Milling equipment, beads, and pottery are largely absent from these loci. Subsurface returns have varied, with debitage counts ranging from 5 to 1,260 pieces per cubic meter. Formal tool counts from subsurface contexts are very low in comparison to other classes of loci. These loci vary in area from small (~ 20 m^2) lithic scatters (e.g., TUL-889B) to extensive (36,625 m^2) flaked-stone workshops (e.g., TUL-896) and seem to represent production, finishing, and repair of stone tools. Such activities often took place in association with small task groups principally focused on hunting upland game.

Territoriality, Boundaries, and Cultural Evolution

Until now the discussion has largely focused on evaluating the form and age of various archaeological loci in the study area and on general changes in land-use patterns over time. However, thus far, the loci have been examined as a unified body of data. Yet prior research has suggested that it may be possible to differentiate study area loci, separating them into groups thought to represent Tubatulabal (n = 53) and Numic (n = 43) (or possibly pre-Numic groups) affiliated populations (Garfinkel et al. 1980: 80–96; Garfinkel 1982; McGuire and Garfinkel 1980:61–69; McGuire et al. 1982; Moratto 1984; TCR 1984:175).

This research first developed categories of site loci independently using certain archaeological data (as discussed above). These constituent site elements (stone tool material, artifact form, rock art style) are recognized as nonrandomly distributed in space and as correlating with defined environmental zones. These archaeological patterns also coincide spatially with territories ascribed to different ethnolinguistic groups; and since that is the case, we might conclude that these ethnic groups and their ancestors are associated with the respective archaeological patterns.

Previous Kern Plateau studies supported just such a division of sites into those located on the Sierra Crest, that were argued to represent Numic (and perhaps pre-Numic) groups, and those in the interior, suggested to be of Tubatulabal origin. Several researchers have accepted these conclusions (TCR 1984;

Moratto 1984), yet others have disagreed (Bard et al. 1985:15–4). Bard and his colleagues suggest that the archaeological patterning may reflect differing adaptive strategies corresponding to varying environmental conditions rather than ethnic/linguistic groups. Bard suggests that the ethnic-affiliation view ignored more significant issues relating to prehistoric changes in adaptive strategies responding to environmental shifts, population movements, and technological innovations.

A look at work conducted almost two decades ago leads to the conclusion that Bard and his colleagues have a point and that their critique merits further consideration. A more integrated and balanced perspective allows today's researchers to use the prehistoric record to posit population associations, yet still pay due attention to changes in the archaeological record that would merit explanation. Such a reconstruction would then focus on the evolution of hypothesized Tubatulabal, Numic, and pre-Numic adaptations. Changing adaptations, environmental shifts, and population movements should also be studied. The present discussion updates prior treatments of the archaeological remains in the study area in light of this more integrated and balanced perspective.

Crestal Versus Interior Site Loci: Settlement Types, Distributions, and Dating

Prior research supports a differentiation of sites in the study area into two groups: those found on and near the Sierra Crest and those located to the west in the interior areas of the Kern Plateau. This distinction is based on: ethnographic data pertaining to the territories of the historic groups occupying these areas, the constituents of the prehistoric sites (including toolstone material composition, milling tool form and composition), style and distribution of rock art sites, and settlement patterns. Prior analysis in the study area did not provide a detailed evaluation of the relative ages and types of loci found within the inventory of sites. These data sets have also not been examined in light of a more expanded site inventory to see if the previous patterns would be supported. Therefore, this study examines the types of loci represented with respect to their relative ages and in reference to the differing archaeological patterns previously documented.

Sierra Crest and Interior Kern Plateau Loci

Tables 5.7 and 5.8 show the frequencies and ages of loci types in the Sierra Crest area and the Interior. Forty-six loci in the Sierra Crest zone and 60 in the Interior are considered.

Environments: Before examining the archaeological patterns in these two distinctive areas, mention must be made of the different environments where these settlements are situated. Crestal settlements are found mainly on ridges and saddles with dense stands of piñon-juniper woodland and sagebrush scrub. These localities are mostly semi-arid and generally devoid of standing surface water. In some cases they have limited areas of level land. Interior localities are much more expansive, occupying open flats, meadows, and river terraces, yet still covered by or associated with nearby dense piñon-juniper forests and situated adjacent to or within easy reach of permanent water sources. Despite these differences, both areas contain abundant stands of piñon pines and have suitably level terrain to allow for temporary settlements during the seasonal harvest of piñon nuts. Chapter 2 contains more detailed discussions of the study area's environments.

Crestal Settlements

Crestal sites have been suggested to be mostly Numic in affiliation (Garfinkel et al. 1980; McGuire and Garfinkel 1980; Moratto 1984). The consensus, held by many eastern-California prehistorians, is that

Table 5.7 Site Loci by Period and Type (Crestal Loci)

Age (years BP)		Classification					
		1 RR	2 MM	3 MNM	4 LS	Subtotal	Percent
Single Period							
Chimney	<650	7	2	2	1	12	**40**
Sawtooth	650–1350	4	3	2	3	12	**40**
Canebrake	1350–3500	0	1	2	3	6	**20**
Lamont	3500–8500	0	0	0	0	0	0
Kennedy	8500–13500	0	0	0	0	0	0
Subtotal		*11*	*6*	*6*	*7*	*30*	*100*
Multiple Period		1	2	2	1	6	
Insufficient Data		7	0	1	2	10	
Total		*19*	*8*	*9*	*10*	*46*	
Percentages		**41**	**17**	**20**	**22**		
Single Period Loci Percentages							
Chimney	<650	58	17	17	8		100
Sawtooth	650–1350	33	25	17	25		100
Canebrake	1350–3500	0	17	33	50		100
Lamont	3500–8500	0	0	0	0		
Kennedy	8500–13500	0	0	0	0		

KEY: Class 1-RR — loci containing rock rings; Class 2-MM — loci with milling (bedrock or portable) implements and midden; Class 3-MNM — components with milling and no midden; Class 4-LS — flaked-stone scatters.

Table 5.8 Site Loci by Period and Type (Interior Kern Plateau Sites)

Age (years BP)		Classification					
		1 RR	2 MM	3 MNM	4 LS	Subtotal	Percent
Single Period							
Chimney	<650	0	5	5	1	11	**35**
Sawtooth	650–1350	2	6	1	2	11	**35**
Canebrake	1350–3500	0	2	1	2	5	**16**
Lamont	3500–8500	0	0	2	1	3	**10**
Kennedy	8500–13500	0	0	1	0	1	**3**
Subtotal		*2*	*13*	*10*	*6*	*31*	*100*
Percentages		**6**	**42**	**32**	**19**		
Multiple Period		0	8	4	3	15	
Insufficient Data		2	1	2	9	14	
Total		*4*	*22*	*16*	*18*	*60*	
Percentages		7	37	27	30		
Single Period Loci Percentages							
Chimney	<650	0	**45**	45	9		100
Sawtooth	650–1350	18	**54**	9	18		100
Canebrake	1350–3500	0	**40**	20	40		100
Lamont	3500–8500	0	0	75	25		100
Kennedy	8500–13500	0	0	100	0		100

KEY: Class 1-RR — loci containing rock rings; Class 2-MM — loci with milling (bedrock or portable) implements and midden; Class 3-MNM — components with milling and no midden; Class 4-LS — flaked-stone scatters.

an initial Numic presence occurs in the study area no earlier than ca. A.D. 600 (Bettinger 1976, 1989; Delacorte 1994; Gilreath and Hildebrandt 1997). Yet others believe that, since this area is either in or near the hypothesized Numic homeland, a Numic presence might be of greater antiquity (earlier than A.D. 600) than in other areas (Grant et al. 1968; Whitley 1994, 1998). Of course the area could still be the homeland for the Numic — if they migrated to the area ca. A.D. 600. That date still would allow enough time so that three to four centuries later the major radiation of the Numic could have occurred when they spread out into the Great Basin ca. A.D. 1000.

Crestal sites might bear evidence for a population replacement or in-migration, with Numic groups initially colonizing and replacing the pre-Numic populations that had used the area prior to A.D. 600. Alternatively, we may find evidence for a continuum of Numic occupation. Investigation of the prehistoric crestal localities documents the use of more ephemeral occupations and those that are more recent in age than in Interior settlements. Sierra Crest loci exhibit fewer middens of shallower depth than do Interior deposits. Class 1 rock ring sites are almost exclusively crestal phenomena. Twenty-nine such features were identified, and all but four were located along the Sierra Crest. Rock ring use began during the Sawtooth Period and increased over time both in number of loci represented per period and in the total number of such features exhibited during each period.

This observation supports the majority view that in eastern California, intensive green cone piñon pinenut procurement began ca. A.D. 600 (Bettinger 1976, 1994; Hildebrandt and Ruby 2000). The shift in settlement subsistence practices was accompanied by resource intensification that including the development of storage facilities (rock rings) used to cache nuts. This shift signals the characteristic "processor" strategy and probably a Numic presence in the region (Bettinger 1994; Bettinger and Baumhoff 1982).

Earlier Canebrake settlements (dating ca. 1550 B.C.–A.D. 600) have sometimes been assumed to be pre-Numic and may differ in character from later settlements (Bettinger and Baumhoff 1982; Delacorte 1994; Garfinkel et al. 2004). Sierran loci appear to display just such a distinction. Class 4 loci (flaked-stone scatters) are proportionally at their highest in the Canebrake and Sawtooth periods, and a dramatic decrease is noted in the later Chimney Period. These loci, routinely identified by eastern-California prehistorians as "logistical hunting camps," follow a pattern of use that appears to be mirrored in our distribution. That is, a decrease in upland hunting camps has generally been recognized as occurring after ca. A.D. 1000 (Bettinger 1989; Gilreath 1999). No obsidian hydration rims larger than 5.2 microns are represented at crestal loci, nor do settlements contain temporally diagnostic projectile points or beads dating earlier than the Canebrake Period. Thus, only extremely limited use of the Sierran Crest can be discerned prior to ca. 500 B.C., and cultural manifestations dating to the Lamont or Kennedy periods are largely absent.

Interior Settlements

Based on historical linguistic studies, most Interior sites are thought to represent the Tubatulabal. According to linguistic reconstructions, the Tubatulabal had in-place longevity on the order of two to three thousand years (Fowler 1972; Hale 1958; Lamb 1958; Miller et al.1971). If confirmed archaeologically, this longevity represents a longer record of indigenous development, with considerably greater antiquity, than the hypothesized recent Numic influx. Such in-place development would lead archaeologists to expect a pattern of greater occupational stability and more continuous use than that characterizing Sierra Crest settlements.

Occupation of the interior Kern Plateau area is in fact represented earlier in time by components dating to both Lamont and Kennedy periods. Early loci are entirely lacking for Sierran Crest settlements. Residential (Class 2) sites predominate in the Interior, representing 42% of the entire inventory. Interior sites also display a continuous emphasis on residential settlements with a long-term preponderance of Class 2 loci. Habitation loci consistently represent between 40–50% of the entire settlement inventory for the duration of several time periods, covering 2,500 years of prehistory. Several Class 2 multiple-period residential sites (e.g., TUL-488N, TUL-629, TUL-621, TUL-879A) also have chronological markers (obsidian hydration rims, time-diagnostic point types) indicating continuous occupation from the Canebrake through the Chimney periods

An increased emphasis on more temporary settlements, including Class 4 (flake scatters) or hunting camps is recognized during the earliest periods of prehistory dating from the Kennedy and Lamont periods through the Sawtooth Period. Differing representation of hunting related sites may represent an earlier unrelated population (other than the Tubatulabal and their direct ancestors), but such an interpretation is purely speculation given the meager data available for those early periods. Rock ring sites, never significantly represented in the interior, existed only briefly and are few in number, with only two loci identified for the Sawtooth Period.

Territoriality

The archaeological identification of prehistoric territorial boundaries has been a subject of considerable debate (Bettinger 1982; Simpson 1988). Yet it seems evident that, with the right circumstances and data sets, the boundaries between hunter-gatherer groups might be defined.

A territorial boundary has been defined that bisects the study area and runs east-west along the northern edge of the Piute and Scodie Mountains then quickly turns north along the crest of the Sierra (Figure 5.1). This line purportedly separates the largely desert-dwelling Numic (Scodie Mountains-Indian Wells Valley territory) from the interior population of Tubatulabal (Kern River territory) (Garfinkel et al. 1980; McGuire and Garfinkel 1980:61–69). If the Sierra Crest acted as a sociopolitical boundary between Numic and Tubatulabal populations, that division represents, in part, the traditional habits and patterning of these populations as represented by their archaeological remains in the study area. This demarcation would most likely represent what has been termed a territorial band or hunter-gatherer district (Steward 1938).

Chapter 3 observed that several different territorial districts did exist for the Tubatulabal and the Panamint Shoshone/Kawaiisu. It was evident that varying degrees of amity-enmity characterized these historic groups. Most importantly, an affinity appeared to exist between the Numic populations of Kawaiisu and Panamint Shoshone speakers, as well as a distinct adversarial stance between Tubatulabal and Numic groups (Irwin 1980; Steward 1938; Voegelin 1938). Members of these hunter-gatherer band districts repetitively used a traditional territory that was largely dictated by the locations of reliable sources of water and a particular set of notable geographic features. The material traces of such territories might be detected archaeologically through the restricted distribution of exotic rock sources. A persuasive argument for territorial distinctions can be made when stylistic marker elements coincide spatially with exotic rock sources on the same landscape. It is fortunate that the study area contains just such an overlapping distribution. Artifacts manufactured of several exotic rock sources spatially coincide with stylistic-marker rock art elements providing strong support for ethnic signatures. This pattern allows this study to posit a boundary based on the spatial patterns of these rock sources and distinctive rock art styles across the landscape.

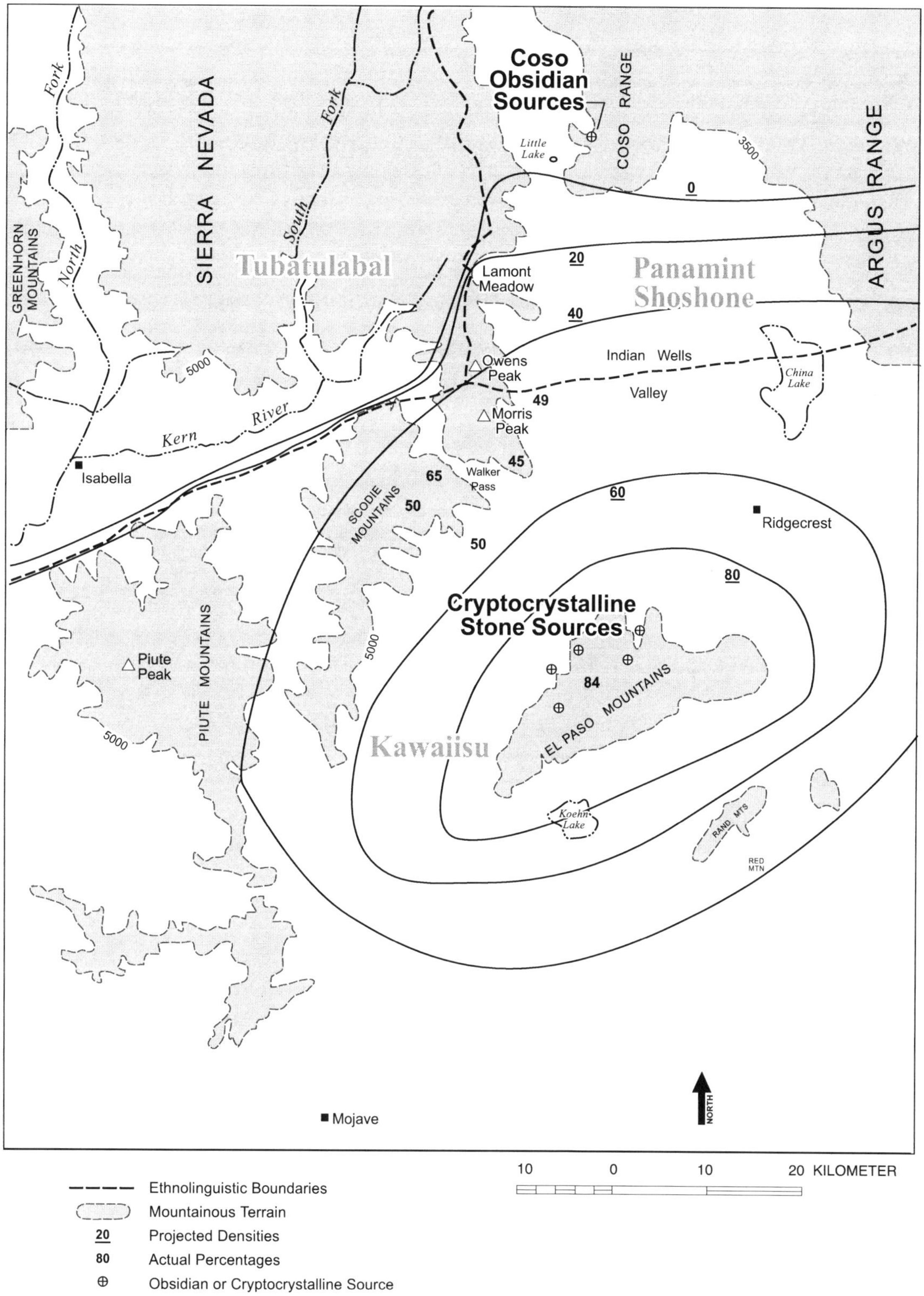

Figure 5.1 Percentage of Cryptocrystalline Materials in Flaked-Stone Assemblages.

If an archaeological boundary can be recognized, a high frequency of a particular type of toolstone might also be expected especially when that material was obtained freely by direct access or through preferential trading relationships. If access were to some degree restricted, as a function of indirect access or by some measure of territorial control, then we would expect a sharp decline for that toolstone. At the edge of a territory (at the boundary) a sharp drop in the frequency of materials would be expected. Mapping of marker artifact materials would show high-density plateaus with sharp well-defined double shoulders and would not display a simple, density distance-decay pattern. Ethnographic information and research with contemporary hunter-gatherers indicates that population increases from in-migration provide circumstances where territorial boundaries are firmed up and may be more marked (Bettinger 1982; Simpson 1988).

Distance Limitations on Contrasting Subsistence Territories

Previous discussions (Garfinkel et al. 1980; McGuire and Garfinkel 1980:63) have recognized that Crestal settlements would have been most effectively used by peoples who occupied winter villages in the desert canyons along the eastern face of the far southern Sierra. Numic peoples in eastern scarp villages would have been able to closely monitor the available subsistence resources in the immediately adjacent Sierran upland piñon environment. Such residential tethering would have allowed for a shortened foraging radius rather than more lengthy and extended forays into the distant upland piñon zones of the Coso Range farther to the west. Interior Kern Plateau encampments, in contrast, would have served as piñon and hunting camp outposts within the seasonal round of the semipermanent Tubatulabal groups who occupied hamlets situated in the South Fork Valley along the Kern River (McGuire and Garfinkel 1980:63, Map 4).

Toolstone Materials

Several types of exotic stone are represented in the archaeological assemblages within the study area. All emanate from the desert to the east, are represented in varying quantities, and have distinctive distributions. Lithic source analysis provides us with a means of mapping the prehistoric behavior patterns of the aboriginal occupants of the Kern Plateau study area and vicinity (Figure 5.1).

Sources and Distribution

The bulk of the toolstone from most loci was acquired from the Coso obsidian quarries 25–40 km to the northeast. Yet crestal loci display much less obsidian and far greater amounts of cryptocrystalline materials than do interior localities (McGuire and Garfinkel 1980:66). The crestal deposits display toolstone assemblages composed of from 5 to 100% cryptocrystalline materials. Visual examination and knowledge of the spatial distribution of toolstone material sources provides strong confirmation that these stone materials originate in the El Paso Mountains. The El Paso Mountains contain beds of agate, opalite, chalcedony, jasper, and petrified wood associated with aboriginal quarries (Davis 1978; Strong 1971:16–21). These cryptocrystalline toolstone source localities are of equal distance from study loci as the Coso obsidian sources are from these same loci, lying only 25 to 40 km to the southeast.

Yet Coso obsidian is the nearly exclusive source (95 per cent or greater) of flaked-stone materials for sites situated within the former territory of the Tubatulabal. Some of these sites are only 0.5 km distant from the Sierra Crest. The nearly exclusive use of Coso obsidian characterizes all interior Kern Plateau study loci and is found far inland at two "hamlet" sites of the Tubatulabal. One of those sites is near the south fork of the Kern River at *ho-lit,* and the other in Long Canyon in the Isabella Basin (KER-311). Both are more than 50 km west of the Coso obsidian sources (Cuevas 2002; Salzman 1977; Schiffman 1980).

The pattern of increased usage of cryptocrystalline stone is also characteristic of the lowland desert areas to the east. A 50/50 split of obsidian and cryptocrystalline debitage is seen at the Freeman Springs site, just south of Walker Pass in the Indian Wells Valley (Williams 2004). An 80/20 division (cryptocrystalline/obsidian) at the Bickel site (KER-250), located in Last Chance Canyon in the El Paso Mountains, also fits the expected pattern (Figure 5.1). Such a distribution is not surprising given the latter site's proximity to the beds of high-quality chalcedony and agate (McGuire et al. 1982). Yet the fact that all projectile points discovered at the Bickel site and a small, yet significant, portion of the debitage are still composed of Coso obsidian reinforces the position that volcanic glass was the stone preferred by many aboriginal groups for their hunting equipment.

An increased reliance on cryptocrystalline materials for toolstone would seem reasonable given the embedded procurement of this material during seasonal use of the El Paso Mountains by Numic populations. The hypothesized Numic group(s) occupying the southern Indian Wells and Fremont Valleys appear to have had only limited and indirect access to Coso obsidian (probably via trade). Otherwise, a greater quantity of obsidian flaked stone would be represented at these aboriginal encampments, as is the case in the territory of the Tubatulabal. The habitual occupation by a territorial band of the Panamint/Kawaiisu would be the most likely purveyors of this pattern.

In contrast, I suggest that the Tubatulabal either had direct access to the Coso obsidian quarries or preferential trading arrangements with the aboriginal group(s) controlling these sources. The former alternative is more likely and may be suggested by Steward's (1938) early determination of a territorial boundary for the Tubatulabal that included the Coso quarry within their homeland.

Groundstone — Material, Sources, Distribution, and Age

Nine localities along the Sierra Crest yielded 16 milling tools, or fragments thereof, manufactured from stone exotic to the Kern Plateau and found only in the desert areas to the east (Table 5.9). None of the 60 interior loci contained any milling artifacts manufactured from exotic volcanic or sedimentary materials. At the latter sites milling equipment is always manufactured from locally available granite. The crestal sites, containing exotic milling equipment, date principally to the Chimney and Sawtooth periods. This pattern of groundstone use is most likely an expression of the habitual use of upland piñon areas by desert-dwellers beginning ca. A.D. 600. This distribution correlates closely with the postulated influx of Numic populations into the area and supports research favoring a late period intensification of piñon exploitation beginning at the inception of the Sawtooth Period (ca. A.D. 600).

Rock Art Styles

Stylistic and locational differences in the rock paintings of the far southern Sierra and eastern California provide further evidence for ethnic distinctions. Rock art sites may have served in part as "stylistically encoded messages of group affiliation and territory" (McGuire 1989; Weissner 1983; Wobst 1974). Such messages have been suggested to be a low-cost strategy for groups to maintain social integration, group solidarity, and territorial boundaries (McGuire 1989).

Morwood (1992:4) also supports the notion that rock art boundaries reflect both geography and the nature of group relations. Thus "the distinctive Wandjina (rock art) style of the Western Kimberleys (in Australia) coincided with the extent of the *wunan* exchange system between linguistically related and culturally similar groups," while abrupt changes in rock art reflected boundaries between aboriginal groups that were traditionally hostile to one another.

Table 5.9 Exotic Groundstone Materials From Sierra Crest Loci.

Site Loci	Description	Dating/Temporal Components
KER-1269	1 Greenish-grey fire-affected metate of volcanic rock	Sawtooth/Chimney
KER-1297	2 Pumice fragments	Multiple
KER-1273	4 Scoria fragments	Chimney
KER-1276A	1 Scoria fragment	Chimney
KER-1276B	1 Sedimentary rock fragment	Chimney
KER-748A	1 Bowl fragment of black vesicular basalt	Chimney
TUL-483	3 Mano fragments of black scoria; 1 fragment of sandstone	Sawtooth/Chimney
TUL-482	1 Bowl fragment of sandstone	Chimney
Isolate collected between TUL-482 and TUL-481	1 Mano of black scoria	Sawtooth/Chimney

Weissner's work (1983) with the Kalahari San echoes such patterning. Her study with the stylistic elements of bone arrowheads allowed her to identify division boundaries between the language groups of the !Kung and G/wi populations (Weissner 1983). Contrasting iconography hints at fundamentally different worldviews and belief systems. Therefore, the coincidence of rock art styles with exotic stone source patterns supports the view that these data fit with ethnolinguistic areas (cf. Quinlan and Woody 2003; Simpson 1988; Wiessner 1983).

Tubatulabal Style

Rock art in the far southern Sierra Nevada differs radically from the pecked designs found on the basaltic boulders and canyon faces of the Coso Range just a few miles east (Grant et al. 1968:108; Heizer and Clewlow 1973; Schaafsma 1986). All recorded rock art sites, except one, are painted rather than pecked (Schiffman and Andrews 1979). Little influence is shown from the Great Basin to the east of the Sierra (Grant et al. 1968). The common element motifs include spoked wheels, peltlike figures, semicircles, rayed circles, and stick figures (sometimes phallic) painted with one color or less commonly multiple color outlines (Table 5.10) (Andrews 1977; Grant et al. 1968; Whitley 1982:158).

Whitley (1982) has statistically correlated element types identified at sites of this style, including those identified as concentric circles, chains, sunbursts, rayed simple circles, rayed concentric circles, and spoked circles. He considers such a correlation as equivalent to a "Tubatulabal Painted Style." The variant he identifies would fit in the Southern Sierra Painted Style, originally identified by Heizer and Clewlow (1973). Whitley demonstrated the validity of this style through his analysis of 1,523 rock art elements coming from 89 sites in the far southern Sierra Nevada. Lee and Hyder (1991) further documented the Tubatulabal variant of the Southern Sierra Painted Style and differentiated it from other neighboring pictograph styles, including patterns evidently associated with the Yokuts, Chumash, and Kawaiisu.

Table 5.10 Artistic Conventions, Pigments, Subject Matter, and Dating for Ethnic Groups.

Style Name	Coso Painted	Southern Sierra Painted
Linguistic Tag	(Numic) Kawaiisu/Panamint Shoshone	Tubatulabalic
Pictographs	X	X
Red	X	X
Black	X	
White	X	X
Yellow	X	
Green	X	
Methods and Settings:	Use of cavities, isolated shelters, granitic and nonbasaltic boulders	Color outlining, open settings and concealed areas
	Dot or dash technique for color outlining	Bilateral symmetry; large scale
Geometric Forms:	Circles, sunbursts, concentric circles, zigzags, vertical dashes, double triangle, linked circles	Circles, sunbursts, dashes, paired half-circles.
Zoomorphic Forms:	Snake, bearpaw (?), bighorn, coyote, cattle, mt. lion, deer, horse	Snake, bearpaw (?), ring-tailed cat (?)
Anthropomorphic Forms:	Stick figures, split heads, lunate-pectoral-like designs, bow-and-arrow hunters	Round-headed figures, split-head stick figures, lunate- pectoral-like design
Other Forms:	Bug-like, pelt figures	Pelt figures
Unique forms:	Painted bighorn sheep, horses, cattle, wide-brimmed hatted anthropomorphs	Ring-tailed cat(?)
Dating	Historic, 100–60 BP (A. D. 1850–post-contact)	Late prehistoric,150–2000 BP (AD 1–contact)

NOTE: Table adapted from Lee and Hyder (1991).

Rock art sites conforming to the Tubatulabal style are found within the study area and are within 0.5 km west of the Sierra Nevada crest. No other rock art style is known within the territory of the Tubatulabal. As to the age for these images, no means to directly date them has been developed. Based on contextual associations and the dates suggested by other researchers (Heizer and Clewlow 1973), they may be estimated to be no older than 2000 years.

Numic Style: Coso Style Pictographs

Garfinkel (1978) first described pictographs of the Coso Painted Style. Two sites at the head of Indian Wells Canyon in the far southern high Sierra were the first sites noted to have similarities in style and subject matter with the Coso Range Representational Style petroglyphs (Grant et al. 1968; Schaafsma 1986). Further work expanded the number of sites manifesting this style (Andrews 1977; Brook et al. 1977; Marcom 2002). Independent evaluation established the style's validity through statistical correlation of element types (Whitley 1982:108–109), supporting a historic age attributed due to the strong correlation of horse and rider elements with bighorn sheep. Researchers collaborated in an anthology focusing on the style, and synthesized what was then known concerning these images (Schiffman et al. 1982).

Style and Subject Matter: The hallmarks of this style are rather elaborate (often polychrome) paintings that nearly always contain images of bighorn sheep and often depict historic Euroamerican subjects (Tables 5.10 and 5.11). Other elements include: concentric circles, handprints, shieldlike patterns, stylized anthropomorphs, deer, hunters with bow and arrows, coyotes, mountain lions or dogs, sunburst symbols, atlatl and dart- or arrow-impaled animals, horses, horse and riders, people with broadbrimmed hats, and longhorn cattle. The paintings contain elements reminiscent of Coso Representational Style Petroglyphs (Schaafsma 1986). The bighorn sheep images often have front-facing, bifurcated horns (a unique feature of the Late and Transitional Period, Coso Range Petroglyph Style). It has been argued that the use of white pigment, the representation of concentric circles, and the presence of handprints are closely associated with Numic Ghost Dance iconography (Caroll et al. 2004; Stoffle et al. 2000).

Geographic Distribution: Coso Painted Style sites are concentrated along the eastern scarp and crest of the far southern Sierra Nevada, in the southern Panamint Range, and in Greenwater Canyon (Marcom 2002:21). Paintings containing such characteristic imagery have now been identified at 20 distinct locations (Figure 5.2; Table 5.11). Coso Style pictographs are found just west of the crest of the Sierra, along the westernmost boundary of Tulare County. Coso paintings are also noted immediately east of the Sierra crest at the head of Indian Wells Canyon and at other locations along the eastern scarp. Other painted panels in the Coso Style are located in the Coso Range, at the head of Surprise Canyon above Panamint Valley, in the Tehachapis at Sand Canyon, and, as their easternmost expression, in Greenwater Canyon, above Greenwater Valley, in the Greenwater Range. Significantly, the nearby Owens Valley apparently contains no painted quadrupeds (zoomorphs or bighorn sheep) and only a few anthropomorphs, and the paintings there are rendered only in monochromatic red pigment (Smith and Lee 2001). The distribution of the Coso Style paintings is coterminous with portions of the ethnographic territories of the Kawaiisu and Panamint Shoshone. Coso Style paintings fall mostly at the interface or borders of the territories traditionally inhabited by these groups. The Coso Painted Style sites have a distinctive, nonoverlapping, discontinuous distribution with the style area for the Tubatulabal variant of the Southern Sierra Painted Style.

The largest and most elaborate Coso Style pictograph (INY-1378) is located in Panamint City. The central panel depicts bighorn sheep and other animals. Some of these creatures appear to be impaled by what seem to be atlatl darts. In two instances there are circles bisected by lines that bear close similarity to renderings identified in the Coso Range petroglyphs confidently identified as atlatls (Grant et al. 1968). These painted elements include conventionalized images of atlatls with finger grips rendered in a fashion quite similar to those represented in the Coso Range petroglyph tradition (Brook et al. 1977:19, Figure 18). Yet the paintings also contain a number of horse and riders and individuals wearing Western-style, wide-brimmed hats (Brook et al 1977; Ritter et al. 1982). A revitalization and reemphasis on traditional imagery would be inferred since atlatls are not known to have been a part of the cultural repertoire at this historic date.

Table 5.11 Characteristics of Sites With Coso-Style Pictographs.

Site Number	Name and Location	Element #	Colors	Element Forms
KER-735	Indian Wells Canyon	41	Red, white and black	Bighorn, horse and rider, cattle, concentric circles, disks, flower-form
KER-736	Indian Wells Canyon	20	Red, orange, pink, white	Bighorn, anthropomorphs, shields, circles, horse and rider, chain
TUL- 478	Lamont Meadow/ S. Crest	16	Black, pink, white, red and orange	Bighorn, geometric, horse and rider
TUL-479	Lamont Meadow/ S. Crest	1	Red	Bighorn
Wasp Nest Cave	Sand Canyon, Inyo Co.	19	Red, white, orange, gray and black	Bighorn, shields, anthropomorphs, geometric
Little Pet. Canyon	Lower Renegade Canyon, Cosos	10+	Red and black	Bighorn, anthropomorph
Day of Freedom	Wilson Canyon, Coso Range	2+	Red and white	Rakes, circles and bighorn
Bierman Caves, SBD-10	Robbers Mountains, South Base, China Lake	15+	Red, black and white	Bighorn, anthropomorph, geometric
Ayers Rock, INY-134	6 miles NW of Coso Hot Springs	65+	Red, orange, blue, black and white	Deer, zoomorphs, bighorn, anthropomorphs, handprints
INY-3250	Trail Canyon, Death Valley	2	Red	Bighorn and anthropomorph

NOTE: KER-735 and KER-736 are discussed in Andrews (1980), Backes (2004), Garfinkel (1978, 1982) and T. Whitley (1982a, 1982b). TUL-478 and TUL-479 are treated in Andrews (1980) and Garfinkel (1978). Wasp Nest Cave is described in Whitley et al. (1982). Descriptions of Little Petroglyph Canyon and Day of Freedom sites can be found in Schiffman et al. (1982). The Bierman Caves have been noted only in a personal communication from Russ Kaldenberg, China Lake Naval Air Weapons Station archaeologist, in 2004. Ayers Rock (INY-134) is illustrated and described in Grant et al. (1968) and Whitley et al. (1982). A recent report describes the results of early excavations conducted by the Archaeological Survey Association offering an analysis of these materials (Whitley et al. 2005)

Table 5.11 Characteristics of Sites With Coso-Style Pictographs (continued).

Site Number	Name and Location	Element #	Colors	Element Forms
DEVA 87E-105	Old Crump Flat, Death Valley	4+	Red	Bighorn, circle and oval
INY-1378	Panamint City Shelter, Panamint Valley	153+	White, red, yellow, black, gray	Bighorn, horse and rider, deer, bird, anthropomorph, geometric, bovine
INY-1379	Ten Gallon Hat, Panamint V.	30+	White, black, red	Bighorn, horse and rider, tally line, anthropomorph
DEVA 87E-124	Greenwater Canyon #4, Main and Upper Shelters, Death Valley	65+	Black, white, red	Bighorn, anthropomorph, tally line, concentric circle, long-horn cattle, starburst, horse and rider, arrow-impaled zoomorph, bird, deer
SBR-089	Unnamed Shelter, Death Valley	7	Black	Bighorn and anthropomorph
INY-3280	Johnson Canyon, Death Valley	7	Red and black	Bighorn, horses?, geometric, anthropomorph
INY-1988	Hanaupah Canyon #1, Death Valley	5+	Red and white	Anthropomorph, bighorn, rabbit?, bird
INY-1989	Hanaupah Canyon #2, Death Valley	6	Red	Anthropomorph, circle, horse, bighorn, horse and rider
INY-4836	The Gallery, Death Valley	25+	Red, black, white and yellow	Anthropomorph, lines of connected anthropomorphs, rabbits?, woman with a dress, bighorn, zoomorph, geometric
KER-508	Tomo Kahni, Tehachapis	50+	Red, black, white, yellow	Anthropomorph, snake, bighorn, geometric

NOTE: DEVA 87E-105, -124, INY-3280, INY-1988, INY-1989 and SBR-89 are described and discussed in some detail in Marcom (2002). INY-1378 and INY-1379 are treated in Brook et al. (1977) and Ritter et al. (1982). INY-4836 is illustrated and discussed in Grant et al. (1986) and Marcom (2002). KER-508 is the subject of discussion in Lee (1991), and Sutton (1981, 2001).

The elements at the largest Panamint City site all appear quite fresh with the pigment seemingly smeared on the rock face. This phenomenon is also characteristic of the paintings at the head of Indian Wells Canyon just below the Sierra Crest sites. The paint at both sites can still be flaked off rather easily, even today (Backes 2004). The preponderance of evidence points to an attempt to copy the earlier iconography found in the nearby Coso Range petroglyphs (cf. Schiffman and Andrews 1982; Sutton 1981).

Ethnographic evidence indicates that Native Americans did indeed copy ancient designs and incorporate them into their artistic traditions with little knowledge of the meaning of such designs (Gifford 1936; Haury 1945:70). In fact, it was pointed out to me that Native Americans in Riverside County are still "refreshing" faded pictographs with manufactured (commercial), oil-based paints (Michael Moratto personal communication 2005). Such an interpretation for Coso Style pictographs (e.g., "Numic Ghost Dance paintings") is supported by the fact that the Ghost Dance ideology was emphatically focused on the past (Carroll et al. 2004).It is also documented, that after the forced relocation of Eastern California Indians by American troops in 1863, there was a cautious and gradual return of Native peoples to their former homelands between 1864 and 1865 (McCarthy and Johnson 2002). When the Natives returned, they found their traditional villages destroyed and homelands occupied by ranchers. Hence, instead of their usual lowland occupation sites, Native peoples occupied more secluded areas, rocky refuge camps at the fringes of and high above the alluvial fans of the white settlements (Walton 1992). Such encampments may have been the sites for the production of Coso Style paintings, and such secluded locations were requisite for proper conduct of Ghost Dance ceremonies (Carroll et al. 2002).

During the 1860s, historical accounts mention Ghost Dance-like activities in eastern California initially, influenced by *Wodziwob*, a tribal shaman of the Mono Paiute (McGrath 1987). *Wodziwob* began to preach his messianic vision at piñon harvest festivals and rabbit hunts. In 1869 *Wodziwob* dreamed that a train was coming from the east and that if native peoples performed the Ghost Dance, they could bring back the dead and restore balance to the world (DuBois 1939; Hittman 1973, 1997).

The Ghost Dance had great similarities to the traditional Round Dance, so that it was relatively easy to graft the religious movement onto the native indigenous cultures (Hittman 1973). Several researchers indicate that the Panamint Shoshone were willing participants in the Ghost Dance movements (Kroeber 1925:872; Mooney 1973:802; Schiffman and Andrews 1982; Steward 1938). During the 1860s and early 1870s, Euro-American depredations against the Owens Valley Paiute, Panamint Shoshone, and Kawaiisu took their most dramatic turn, and cultural destruction of their traditional lifeways reached its zenith.

If there is a correlation between the period of cultural destruction and the production of the Coso Style paintings, then rock art sites containing historic elements (horses, horse and riders, anthropomorphs with wide-brimmed hats, bovine elements) and painted sheep might date from no earlier than 1850. The historic depictions of horse and riders and longhorn cattle found within the Coso paintings may be visual records of the unusual and dangerous activities observed and recorded by the native inhabitants of eastern California. Such shattering experiences of colonialism may have fueled a revival in a tradition of rock art as exhibited in the Coso Style Paintings (Garfinkel et al. 2007; Quinlan and Woody 2000).

Ethnic Affiliation: The physical location (Figure 5.2), likely historic and protohistoric dating, subject matter (horses, mounted riders, hatted anthropomorphs, and longhorn cattle) and associated archaeological materials would strongly suggest that the Coso paintings were rendered by the historic native inhabitants of the areas in which these rock art sites are found. It would seem reasonable that the manufacturers of the Coso Style paintings were people who spoke a Numic language. The most likely candidates were Native Americans speaking Kawaiisu and/or Panamint Shoshone.

A number of subgroups or territorial bands of the Panamint Shoshone evidenced a mix of speakers of Kawaiisu and Panamint. Two Panamint Shoshone Districts, the Koso and Panamint Valley Districts, contain most of the known and many of the largest Coso Style paintings. These districts have been described by a number of anthropologists as having a mixture of native peoples. Steward (1938: Figure 7), Driver (1937), and Sennett-Graham (1989: 25) reconstruct the precontact Koso District (*Pawo'nda*) as containing members who spoke Panamint Shoshone but also speakers of Owens Valley Paiute and Kawaiisu. Similarly the Panamint Valley (*Haita*) and southern Death Valley Districts (*Tumbica*) in precontact times are thought to have had an almost equal balance of Shoshone and Kawaiisu with their southernmost portions being predominantly Kawaiisu (Driver 1937; Sennett-Graham 1989:25; Steward 1938: Figure 7).

Cultural Sequence

Since we recognized a sharp definition of a supply and falloff zone for exotic stone (obsidian, cryptocrystalline, volcanics, and sandstone) and distinctive differences in rock art, it seems that it might be possible to date the division between the Indian Wells/Scodie Mountains and Kern River territories. Most evidence for those territories, including the time depth for sites having exotic groundstone and piñon storage caches situated on the Sierra Crest (Table 5.9), suggests an age after 1350 B.P. when occupation is the most intense and the archaeological record most complete. A review of crestal sites shows no correlation by age and percent of cryptocrystalline stone (either for increasing or decreasing composition) or an association by loci (settlement type). Hence the boundary seems to have been one of some prominence for nearly 1500 years (ca. A.D. 600–contact).

Summary

Four classes of archeological components are recognized in the study area. These include rock ring loci or temporary piñon camps where green cones were cached and piñon nut harvests occurred. Residential settlements (larger base camps) are recognized where aboriginal groups overwintered during exceptionally abundant piñon years. Milling stations with bedrock features, and flaked-stone scatters complete the range of loci types enumerated for the Kern Plateau. Flaked stone scatters mostly served as logistical hunting camps or, in exceptional cases, flaked-stone workshops used for the production of Coso obsidian bifaces intended for export.

Changes in land-use patterns over time indicate that most occupation in the region occurred during the last 3500 years, with episodes of especially intensive use occurring over the last 1350 years. Crestal sites tend to be more recent than sites from the Interior. The former localities probably represent the influx of Numic groups into the region beginning ca. A.D. 600 and indicate subsistence intensification related to piñon (green cone) procurement. Interior sites manifest a more continuous occupation from earlier times and consistently display more permanent residential occupations probably affiliated with Tubatulabal populations. A boundary between the Kern River and Indian Wells/Scodie Mountains territories is indicated based on disparities in the percentages of exotic stone materials, presence of nonlocal materials for milling tools, distances to winter villages and semipermanent hamlets, and styles of rock paintings. This territorial division appears to have been established by 1350 B.P.

Chapter 6
Linguistic Archaeology

Scope and Purpose

In this chapter the timing and character of prehistoric population movements and *in-situ* developments within the study area are considered. The discussion follows closely earlier treatment (Chapter 1) of test implications for static versus dynamic models of cultural development. In turn, data supporting or refuting continuous cultural traditions or discontinuities in the historic territories of the Tubatulabal and Numic peoples are examined. Alternative models for Numic linguistic prehistory are also evaluated with reference to archaeological data from the Kern Plateau and eastern California.

Evaluation of In-Place Versus Replacement Models

Archaeologists and linguists agree that the Tubatulabal language is longstanding and may result from an unbroken record of local development. Historical linguistic data are not incompatible with an in-place development of Tubatulabal for a minimum of two to three thousand years (Bettinger 2002; Foster1996; Fowler 1972; Lamb 1958; Miller 1986; Moratto 1984). No evidence of population movements, expansions, contractions, or replacements is suggested by language distribution or linguistic criteria. Hence, a good candidate for comparative evaluation is the "Tubatulabal case."

Assuming that linguists have accurately characterized the time depth and territorial stability for the Tubatulabal, this pattern would be distinctively different from that in the southwestern Great Basin. In the latter area, linguists largely agree that population movements did occur, and a recent expansion of Numic populations apparently took place (Foster 1996; Miller et al. 1971). Since the Numic speakers never replaced the Tubatulabal, one might reasonably expect to see few, if any, archaeological manifestations that would characterize the replacement of pre-Numic groups by the entry and expansion of Numic groups (Bettinger and Baumhoff 1982; Elston 1994).

A review of the archaeological record might show what stability would look like (the Tubatulabal case). If the prehistoric remains were consistent with such a pattern, as expected, that consistency would support the hypothesis of continuous *in-situ* development. Examination of the upland Tubatulabal homeland should provide a basis for comparison with the adjacent region of Numic and possibly pre-Numic occupation. Tubatulabal continuity might be seen in a pattern of prehistoric site use that displays relatively continuous, long-term, sustained occupation from the historic era back 2,000 to 3,000 years with no discernible breaks.

Although sometimes difficult to detect archaeologically, certain geographic locations provide unique sets of environmental factors that attract continuing, longterm occupations. *In-situ* occupations showing such longterm use, with no lengthy gaps (periods of abandonment), would support an interpretation of continuous, autochthonous development. Gradual transitions between artifactual assemblages and uniformity in stylistic attributes over time would also denote such continuity. A limited range of rock art style would be representative of a single cultural tradition. A gradually increasing and relatively unchanging obsidian hydration curve, with little to no evidence for changes in access (or territoriality) over time, might also support a relatively stable pattern. Minimal shifts in obsidian source use over time might also be expected.

Artifact types would be predicted to have standard bell curve distributions, beginning slowly with their early introduction, peaking in popularity, and trailing off as their popularity diminished. In some cases, artifact distributions overlap with the subsequent introduction of alternate artifact types that were replacing the waning forms. Of course, a type might also disappear abruptly as a result of a technological change (such as the advent of the bow and arrow) even when a population remains in its territory.

Resolution and Interpretation of Coso Obsidian Hydration Chronologies

Obsidian hydration measurements have been routinely used as an indicator of aboriginal activity throughout California (e.g., Basgall and McGuire 1977; Gilreath and Hildebrandt 1997). Yet it has often been noted that the chronological divisions of the local cultural sequences are commonly of varying lengths, potentially confounding the archaeologist's ability to measure occupational intensity and site use in a simple manner. In order to produce more valid measures of site and regional use over time, some researchers have recommended that these measures be "time-adjusted." In other words, the absolute frequency or relative percentage measures might need to be slightly revised to make the periods directly comparable.

Yet another influential factor is that flaked-stone reduction, in late prehistoric periods (of brief duration), generally produced less debitage than during the earlier (and longer) cultural periods. This relative scarcity of debitage is caused in large measure by the change in technology from heavier dart points to smaller arrow points. However, many more arrow points were produced than dart points since they were lost more easily and the technology of arrow production and use requires a larger supply of arrow points than dart points. Additionally, the flaked-stone debris produced during more recent intervals was usually smaller in size, adding another potential sampling bias. Reworking/ scavenging activities also tends to deplete older assemblages and amplify younger ones.

Nevertheless, the number of hydration readings exhibited per time period is always weighted toward the eras of highest obsidian import, discard, and production. The present evaluation assumes that obsidian hydration readings are a generally valid method, albeit imperfect, to measure prehistoric activity. This discussion uses such measures as one means of monitoring the character and intensity of cultural activity in the study area.

The Tubatulabal Pattern — Evidence for Autochthonous Developments

The Direct Historical Approach

When no evidence for historical discontinuity can be demonstrated, it is often plausible to suggest that the ethnographic cultural groups also operated in the past. Occupation of the Interior Kern Plateau area, hypothesized to be Tubatulabal, is represented by a number of upland residential settlements (TUL-488N, TUL-629, TUL-621, TUL-879A). These sites manifest relatively continuous occupation over the last 3,000 years, with no evident breaks in their cultural sequences. Additionally, TUL-488N is located at the edge of Chimney Creek Meadow in an area that is consistently identified in ethnographic and ethnohistoric sources as having been one of the main traditional pinenut grounds for the Tubatulabal (Butterbrett 1948; Powers 1971, 1981; Voegelin 1938).

A number of multiple-period residential settlements (TUL-488N, TUL-629, TUL-621, TUL-879A) exhibit obsidian hydration rim frequencies attributable to several time periods from Canebrake through Sawtooth and continuing through the late prehistoric era into the Chimney Period (Table 4.3). Two of these sites (TUL-629 and TUL-621) produced evidence of continuous occupation from prehistoric times directly transitioning to the historic era. Historic Mission Period European glass trade beads were recovered from excavations at TUL-629 (n = 18) and TUL-621 (n = 6) and a continuous suite of hydration rims ranging from about 1.0 to 5.2 microns is exhibited at both sites (Table 4.3). These deposits appear to have been regularly occupied from about 2400 years ago to the historic era.

Cultural Sequences: Site and Regional Chronologies

The Lake Isabella Basin is the geographical setting for the lowland hamlets of the ethnographic Tubatulabal. Dillon (1988) reports hydration values on 50 obsidian specimens from seven archaeological sites in this area. Sutton et al. (1994) present 22 additional readings from seven sites, also in the Lake Isabella Basin.

Gehr (1981, 1988) developed a mean temperature gradient used in refining the Coso obsidian hydration rate and correlated 13.59 degrees centigrade with an elevation of 835 m (3,040 feet amsl). Lake Isabella is located at an elevation of approximately 800 m (2,600 feet amsl), nearly the same as Gehr's estimate. Therefore, the same correction factor can reasonably be applied to the Isabella hydration rims (cf. Gilreath and Hildebrandt 1997).

The composite hydration curve resulting from the combined Lake Isabella data sets is broadly similar to that of the upland interior Kern Plateau (Table 6.1). Both show a unimodal peak with the highest number of readings centering on the Sawtooth and Canebrake intervals. Yet substantial percentages of the measurements also fall within the Chimney Period. These distributions suggest maximal exploitation of Coso obsidian and perhaps increased levels of aboriginal activity during these intervals.

The Lake Isabella Basin and Interior Kern Plateau rim distribution suggests a simple normal distribution or bell curve, with continuous activity over time manifesting little in the way of disruption or upheaval. An uninterrupted sequence is shown with a significant proportion (20% or greater) of the hydration rims in three consecutive temporal periods on the Kern Plateau and a largely similar sequence represented for the Lake Isabella Basin (Table 6.1).

The obsidian hydration curves for the Isabella Basin and Interior Kern Plateau (Table 6.1) are closely concordant and show intensive use consistent with linguistic estimates for an *in-situ* Tubatulabal development on the order of 2,000-3,000 years (Hale 1958; Lamb 1958). Such a relatively homogeneous distribution is entirely consistent with an in-place, continuous development of the Tubatulabal within their hypothesized homeland.

A review of the obsidian hydration rim tabulations from the Sierra Crest and the desert region reveals significant differences (Table 6.2). These readings display a more punctuated distribution with abrupt discontinuities, implying periodic site abandonment and episodes of site disuse.

Lake Isabella Basin Sites

Sequence	OH Rims in microns	Time Span (years B.P.)	N	%
Chimney	<3.7	< 650	12	16.9
Sawtooth	3.7-4.9	650-1,350	19	26.7
Canebrake	4.9-7.6	1,350-3,500	29	40.8
Lamont	7.6-16.0	3,500-8,500	11	15.5
Kennedy	16.0-21.1	8,500-13,500	0	0
		Total	**71**	

Interior Kern Plateau Sites

Sequence	OH Rims in microns	Time Span (years B.P.)	N	%
Chimney	<2.4	< 650	72	23.2
Sawtooth	2.4-3.7	650-1,350	105	33.9
Canebrake	3.7-6.5	1,350-3,500	99	31.9
Lamont	6.5-10.1	3,500-8,500	30	9.6
Kennedy	10.1-13.9	8,500-13,500	4	1.3
		Total	**310**	

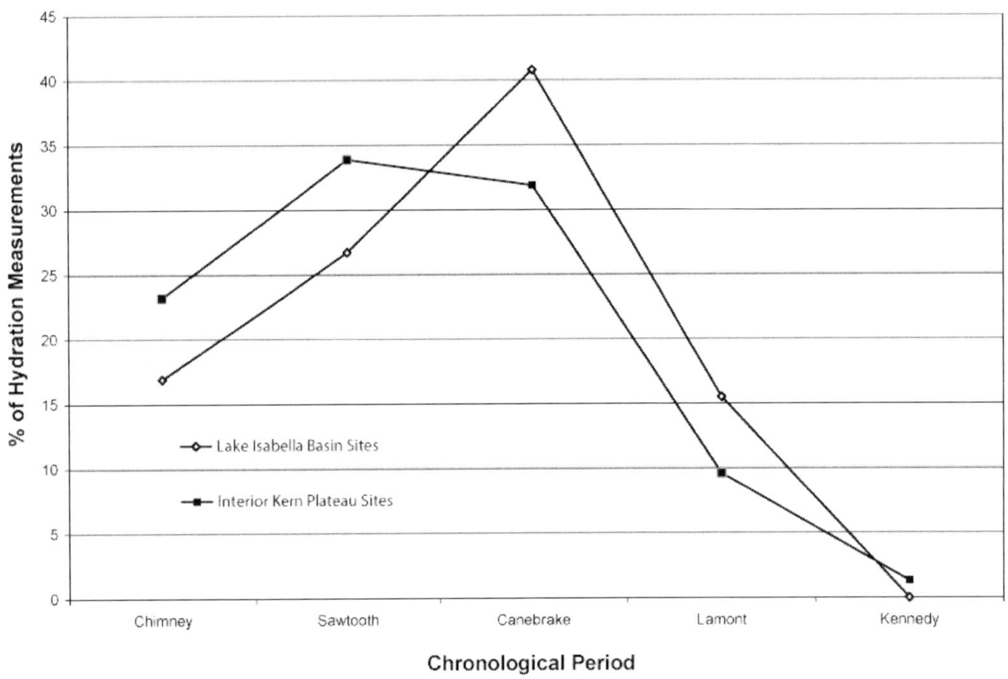

Table 6.1 Summary of Hydration Readings on Coso Obsidian From Lake Isabella and Interior Kern Plateau.

Table 6.2 Distribution of Coso Obsidian Hydration Rims for Haiwee/Marana and Chimney/Sawtooth Assemblages in Eastern California.

Lowland Coso Obsidian Hydration Rims*	<3.7 Marana	%	3.8–4.5 Late Haiwee	%	4.6–5.2 Early Haiwee	%	5.3–7.6 Newberry	%	7.7–16.0 Little Lake	%	Totals
Ash Creek (Iny 3812)	1	1.75	32	56.14	22	38.60	2	3.51	0	0.00	23
Bickel (Ker 250)	3	8.82	22	64.71	3	8.82	6	17.65	0	0.00	57
Maggie's Site	5	21.70	13	56.52	2	8.70	3	11.50	0	0.00	23
Junction Ranch (Iny 1534B)	32	50.00	17	26.56	5	7.81	5	7.81	5	7.81	64
Upland Coso Obsidian Hydration Rims	<2.4 Chimney		2.5–3.0 Late Sawtooth	%	3.1–3.7 Early Sawtooth	%	3.7–6.6 Canebrake	%	6.7–10.7 Lamont	%	
Sierra Nevada Crest	34	24.11	27	19.15	41	29.08	38	26.95	1	0.71	141
	Marana		Late Haiwee		Early Haiwee		Newberry		Little Lake		
	<2.1		2.2-2.7		2.8-3.2		3.3-5.3		5.4-7.0		
Coso Piñon Forest	44	24.31	15	8.29	15	8.29	74	40.88	17	9.39	181

NOTE: Readings for the Ash Creek site are from Gilreath and Holanda (2000). Bickel site readings are from McGuire et al. (1982). Junction Ranch hydration rims are excerpted from Hillebrand (1972). Kern Plateau Crest sites are a summation of data presented here in this dissertation (see Chapter 4 for more detail). CosoPiñon Forest site data are from Hildebrandt and Ruby (2003). Maggie's Site data are from Garfinkel et al. (2004), outliers excluded. Percentage distributions for chronological periods of about 20.00% are underlined in bold to emphasize substantial activity.

*Hydration readings included are for sites with hydration rim measurements on Coso obsidian situated at elevations from 3000 to 5000 feet amsl and using the lowland Coso obsidian hydration rate parameters suggested in Chapter 5 of this dissertation.

Subsistence-Settlement Regimens

Site TUL-488N (Chimney Creek) yielded faunal remains (Basgall and Hildebrandt 1980; Garfinkel et al. 1980: Table 38) that support a consistent and predominant reliance on large mammal exploitation throughout the entire occupation span. Large mammal bone counts predominate in almost every level of every excavation unit at the site, and artiodactyl exploitation was emphasized throughout the entire occupation (Basgall and Hildebrandt 1980:308). This pattern is remarkable considering that site use spans at least 2500 years from the Canebrake through Chimney periods. Little in the way of subsistence change and a remarkable degree of continuity with respect to hunting patterns are represented in the deposit.

Similarly, piñon pinenut shells, piñon pine cone scales, and grey pinenut fragments were recovered throughout the deposits from the uppermost to the lowest excavation levels at the four principal upland residential settlements in the interior Kern Plateau (TUL-448N, TUL-629, TUL-621, TUL-879A). Although these charred plant materials were not directly radiocarbon dated, they still suggest a degree of continuity in subsistence pursuits throughout the prehistoric era. These pinenut base camps with their complement of charred pinenut remains support the interpretation that these settlements witnessed regular exploitation of the local pinenut tracts over a very lengthy span of time (cf. Rhode 1980c, 1980d).

Toolstone Access, Exchange Relations, and Territoriality

Source determination of the artifactual obsidian from the interior Kern Plateau indicates a nearly exclusive use of Coso volcanic glass throughout prehistory. Researchers in eastern California have often noted changing frequencies in obsidian source use as a characteristic pattern associated with changes over time in residential mobility and territoriality (Basgall and McGuire 1988). Yet archaeological sites located in the ethnographic territory of the Tubatulabal fail to show any such changes.

Regional trends in overall obsidian source variability for eastern California indicate marked decreases over time in toolstone diversity. Eastern California pre-Newberry and Newberry contexts produce obsidian source profiles and non-obsidian flaked stone commonly interpreted as evidence of an extremely wide-ranging settlement pattern characterized by extensive residential mobility (Basgall 1989; Bettinger 1999; Delacorte 1999). In contrast, mobility patterns for the Tubatulabal must have been fairly stable since no significant representation of non-obsidian flaked-stone or diverse obsidian source profiles has been observed. Yet the adjacent settlements along the Sierra Nevada Crest do manifest just such a regular complement of non-obsidian toolstone (see Chapter 5), a pattern more akin to that of the Numic, and possibly pre-Numic, groups in the desert region to the east.

Style and Cultural Traditions

Only rock art sites conforming to the Tubatulabal Style are found within the territory of the Tubatulabal (see Chapter 5). This regional variant of the Southern Sierra Painted Style is recognized within the ethnographic territory ascribed to the Tubatulabal and is found within the Kern River drainage and interior Kern Plateau. This style of rock art is absent along and within the Sierra Crest area. No distinct differences in location, function, or subject matter have been recognized.

Based on contextual associations and the dates suggested by other researchers (Heizer and Clewlow 1973), these paintings apparently date as early as 2000 B.P. and were produced continuously until the historic era.

Summary

A variety of archeological evidence supports the conclusion that the Tubatulabal language and cultural tradition are long-standing in the area of the South Fork of the Kern River (Isabella Basin) and the interior of the Kern Plateau. The direct historical approach applied to several archaeological sites shows continuous, unbroken occupation from the historic era back 2500 or more years. Distributions of obsidian hydration rim values from the Isabella Basin and interior Kern Plateau bolster the contention that a relatively uninterrupted prehistoric cultural sequence exists in the area. Subsistence data also support this conviction, with a surprising degree of stability in the profile for vertebrate fauna and plant exploitation patterns perhaps over the past 2500 years. Trends in obsidian source use and acquisition and the representation of a single rock art tradition of the Southern Sierra Painted Style (the Tubatulabal variant) further testify to a single cultural expression of long duration (throughout the last several millennia).

The Numic Pattern — Evidence for Late In-Migration and Population Displacement

Population displacement and immigration are often recognized by abrupt changes in the archaeological record. In some instances such patterns have been verified independently using population genetics (cf. Hayes et al. 2002). The introduction of an exotic population may be visible archaeologically and represented by significant and dramatic shifts from a prior pattern and should not be associated with the relative stability or gradual change characterized by the duration of a cultural tradition. Hence, breaks in the cultural sequence may reflect cultural succession and population movements.

If substantial and dramatic population movements took place, on the order of a Numic/pre-Numic replacement (cultural succession), then this would most likely be reflected by distinctive breaks (disconformities) in the regional and site-specific cultural chronologies and obsidian hydration curves. Such pattern changes can be evaluated using regional and site-specific obsidian hydration and artifact chronologies. Widescale disruption in the regional patterning of site occupations could denote population movements, in-migration, expansion, displacements, and/or cultural succession.

The Direct Historical Approach

Considerable evidence exists to posit a direct historical connection between the ethnographic Numic populations in eastern California and late prehistoric archaeological manifestations in this region.

Chapman Cave (INY-1534A): Some of the most persuasive archaeological evidence linking the historic Panamint Shoshone to their late prehistoric archaeological signature is discerned from studies at Chapman Cave (Chapman 1). The rockshelter was named after the father and son who discovered this small shelter in 1965 while hunting on Wild Horse Mesa in the Coso Range. The site lies at the base of a basalt cliff near upper Renegade Canyon. Hillebrand's dissertation (1972) provides the only description of the shelter and the prehistoric materials recovered from it. Gilreath (2000: 9–49) presents a recent reappraisal of the materials.

At Chapman Hillebrand excavated four cache pits, five burials, four hearths and a collection of sunbaked clay bowls and figurines. The number of individuals interred at the shelter argues for the use of the cave as a cemetery. The cache pits contained historical materials including machinemade cloth, wool trousers, and a dynamite percussion cap as well as typical precontact objects such as twined basketry, flaked-stone artifacts (projectile points and debitage), and pigment. The five burials included seven individuals, with two having indications of cremation. The accompanying mortuary offerings included twined basketry, projectile points, other bifaces, a rabbit-skin cloak and assorted cultural materials (Hillebrand 1972; Gilreath 2000).

Three radiocarbon dates indicate site use from at least 775 to 285 B.P. (Gilreath 2000). Nine Desert Series, ten Rose Spring, and two Humboldt points suggest primarily Haiwee and Marana occupations. Cremation was the common burial practice of the historic Death Valley Shoshone (Wallace and Wallace 1978, 1979); yet when fuel was scarce an individual would be buried in a flexed position below a rock cairn. Grosscup (1977:127), working with C. Hart Merriam's notes, documented historic cremations in Death Valley. Yet the Owens Valley Paiute (Steward 1933:296–299) normally buried their dead shrouded and in a flexed position. Cremation was accorded to those who died outside of their home territory as a practical means of returning their remains home (Steward 1933).

Archaeological studies of Death Valley document cremation only during the Marana Period, as opposed to the Haiwee era when flexed, primary inhumations, often covered by a rock cairn, are reported (Wallace at al. 1959). Two of the seven burials from Chapman 1 are cremations and evidently have close affinities to the historically documented, traditional practices for the treatment of the dead by Numic groups in the area.

The basketry from Chapman 1, reviewed by Polanich (2000), includes both twined and coiled fragments, conforming closely in style, construction, materials and technique to known ethnographic Panamint examples. The sunbaked clay miniature vessels and figurines are nearly identical to those recovered by Wallace (1965) at a site near Stovepipe Wells in Death Valley and are assumed to be ritual objects dating to the late prehistoric and historic eras.

Polanich (2000) and Gilreath (2000) posit that Chapman 1 was first occupied ca. A.D. 450 and used to a considerable extent through A.D. 1850. Nevertheless, they argue that only certain burial patterns (cremations), basketry types, sunbaked clay figurines, and miniature clay vessels are significant ethnic markers (Gilreath 2000:30), attesting to the Numic-affiliation of the remains from about A.D. 1300 to historic contact.

However, since Chapman 1 exhibits a *repetitive pattern of traditional use as a burial chamber*, it seems most likely that all the burials (cremations and inhumations) and other materials placed there are part of the same cultural tradition. Hence, the chronological indicators would seem to argue for an unbroken continuum of Numic use of the site, and indirectly the Coso area, from ca. A.D. 450 to contact.

Mitochondrial DNA (mtDNA) and Numic Antiquity

No automatic correlation exists between biology and language. Yet biological evidence has been found to be the most direct source of information concerning prehistoric population distributions and movements. An examination of ancient DNA extracted from prehistoric human remains for evidence of genetic continuity can be used to evaluate models of population replacements and expansions (cf. Eschleman 2002). A landmark series of recent studies by Kaestle and two of her colleagues provides some limited evidence supporting the Numic expansion/replacement hypothesis (Kaestle 1998; Kaestle and Horsburgh 2002; Kaestle and Smith 2001). Kaestle and Smith (2001) argue that mtDNA from prehistoric burials in the western Great Basin is inconsistent with an early occupation of that region by Numic groups.

The mtDNA of modern Native Americans has been shown to fall into one of at least five haplogroups (A, B, C, D, or X) whose frequencies differ among various ethnolinguistic/ tribal groupings. However, modern Numic populations (e.g., Northern Paiute and Shoshone) typically have C (17% on average), a high frequency of D, but no A haplogroups. The frequency distribution of mtDNA haplogroups of the ancient inhabitants of the western Great Basin (burials dated from 6000 to 300 B.P. radiocarbon years uncalibrated and presumably excluding late prehistoric Numic remains) were recently examined in 39 individuals. These skeletal materials were recovered from Stillwater Marsh and Pyramid Lake in western Nevada and provide some intriguing results that might suggest a recent population replacement by contemporary Numic residents within the last 500 years (Kaestle and Smith 2001).

The mtDNA of ancient pre-Numic groups does not appear to resemble that of modern inhabitants of the western Great Basin. These prehistoric archaeological skeletal remains, as a group, contained low frequencies of A (8% on average) and no C haplogroups. Percentages of B and D varied but were rather similar between ancient pre-Numic groups and historic Numic populations. Neither group has been shown to contain X (Kaestle and Smith 2001). A sample size of 39 is quite small for population studies over a time span of more than six millennia. However, this intriguing pattern appears to at least be compatible with models favoring a recent expansion and population replacement by Numic groups in the Great Basin.

Additionally, eight human coprolite (preserved excrement) samples from Fish Slough Cave in the northern Owens Valley provided well-preserved mtDNA (Kemp et al. 2004). One sample is attributed to Haplogroup B, five are members of C, and two are members of D. The B and D samples are not informative as to their genetic population. However, based on stratigraphic and chronological information (Nelson 1999), the coprolites assigned to Haplogroup C might exclude pre-Numic associations, since ancient skeletal remains have been posited as having no such mtDNA. In contrast, historic Numic populations apparently exhibit some (17%) C Haplogroup associations; therefore, the genetic attribution of those samples, which appear to be contemporaneous with strata dated from about A.D. 840 to 1180 (midpoints of calibrated radiocarbon dates), is likely to be Numic.

Chronological indicators at Fish Slough Cave included Cottonwood points, Queen obsidian hydration measurements, and Owens Valley Brownware. These indicators are consistent with an expanded range of occupation for the cave that includes the Marana Period (A.D. 1300–Historic) and runs from ca. A.D. 840 to the historic period. The Haplogroup C samples provide a sequence motif identical to the haplotypes found in contemporary Numic populations (Northern Paiute or Shoshone). Hence, Nelson's data would suggest an initial presence of Numic groups in the northern Owens Valley by no later than A.D. 800, with a continuing presence into historic times. Such an interpretation is not inconsistent with the suggested Numic presence, based on the Chapman Cave data (above), from A.D. 600 through the historic era. Unfortunately, no data are currently available from Fish Slough Cave to make a determination of the genetic population characteristic of the critical period antecedent to the Haiwee interval (before A.D. 600).

These recent analyses of mtDNA are consistent with the position that population replacement occurred within the western Nevada area of the Great Basin (Stillwater Marsh and Pyramid Lake area). A genetically distinct group, different from the historic Numic inhabitants of that area, can possibly be recognized within the earlier prehistoric burial population. The efforts by Kemp et al. provide some limited evidence for Numic antiquity in the northern Owens Valley — again not inconsistent with the suggestion that this may have been one area from which the Numic expansion occurred. Since Numic

DNA is found in this context far earlier than recognized in the Stillwater Marsh/Pyramid Lake samples, the northern Owens Valley may have been one of the homeland areas from which the Numic people dispersed. Alternatively, the valley could be an area occupied early in the dispersal sequence.

Numic Haplogroups in the Kemp study dated as early as A.D. 800, and older diagnostic materials were not found. Remains older than A.D. 600 could represent a pre-Numic occupation, as might be expected if a population replacement (cultural succession) included the Owens Valley as well as other areas of eastern California. Alternatively, the northern Owens Valley groups could have been continuously Numic, still leaving considerable latitude open to a population replacement and pre-Numic occupation of the far southern Sierra Nevada, Rose Valley, and the Coso Volcanic regions.

Chronology, Cultural Sequence, and Subsistence-Settlement Strategy

The far southern Sierra Nevada Crest and eastern California witnessed a significant series of adaptation shifts beginning ca. A.D. 600. During the onset of the Haiwee Period (A.D. 600–1300) a dramatic set of subsistence-settlement changes were documented. These changes include: less large game hunting, increasing reliance on dryland hard seeds, the beginning of intensive green-cone piñon pinenut exploitation, and the development of sites emphasizing the acquisition of easily procured and abundant small game animals (especially with respect to large numbers of lagomorphs and grebes). These cultural changes may reflect a Numic in-migration. These technological innovations and labor-intensive adaptive strategies are also broadly consistent with those of Numic groups (Bettinger and Baumhoff 1982; Delacorte 1994, 1995). Such an adaptation, it has been argued, would have provided Numic peoples with a competitive advantage over existing pre-Numic populations since it would have enabled them to exploit a wider range of resources that were more costly to collect and process. Hence resources with high extractive and processing costs would have been exploited only after the arrival of Numic groups in the area (cf. Bettinger and Baumhoff 1982; Delacorte and McGuire 1993).

The Haiwee Period was also a time when Numic and pre-Numic adaptations may have coexisted in eastern California (see Contemporaneity of Numic and Pre-Numic Patterns, below). Numic pioneers, during their initial presence in eastern California, may have occupied secondary, and from the pre-Numic perspective, inferior settlement locations. Numic sites are different from and perhaps located in more "marginal" resource areas than sites previously established and "reserved" by longtime resident pre-Numic occupants of the area (cf. Delacorte 1994; Garfinkel et al. 2004; Gilreath and Holanda 2000).

From a careful study of the archaeological record, a pattern of lowland, intensive small-game hunting camps appears to have occurred with the development of large-scale, intensive, upland green-cone piñon pinenut exploitation. This pattern also is contemporaneous with an initial focus on the acquisition, mass processing, and storage of dryland seeds (Basgall and Delacorte 2003; Basgall and Giambastiani 1995). These seed camps routinely include rock rings, thought to be the foundations of brush structures. Many of these rock structures contain doorways facing toward the rising sun and are associated with numerous handstones, milling slabs, and bedrock grinding features.

Single-component Haiwee-age hunting camps are frequently located in "geographically isolated areas" (Delacorte 1994). Such localities provided access to a limited range of biotic communities and appear to have a rather specialized focus on a narrow array of subsistence resources. Therefore, these settlements are a distinctly different group of sites from earlier or later occupations that tend to overlap at the same settlements and hence evince a lack of continuity from earlier settlements.

Dryland Seed Camps: The Coso Junction Ranch Site (INY-1534B) is a multi-component site that includes hunting blinds, petroglyphs, and house rings. The house rings are physically differentiated from the other materials, being located on terraces below the other remains at the site. The rock rings are situated in two groups on a basalt flow that slopes to the floor of Etcheron Valley within the Coso Range. The site is about five kilometers northwest of Chapman Cave (discussed above) and lies at an elevation of 4500 feet amsl. Timothy Hillebrand's dissertation (1972) is the only source of information on this site.

The main site expression is a notable grouping of eight house rings. Identified in association with these rings was an extensive scatter of stone flakes and tools, projectile points, sherds of Owens Valley Brownware, 12 handstones, 75 basalt metates (some deeply worn into trough shapes), and five bedrock mortars. Projectile points were exclusively of Marana and Haiwee age and include nine Desert Series (Cottonwood and Desert Side-notched) and 51 Rose Spring forms. Sixty-four obsidian hydration readings were reported in the original study, and all but five fall into a unimodal distribution ranging from 5.6 to 1.6 microns (Table 6.2). Disregarding a small group of outliers (perhaps anomalous early readings), the largest grouping of rims (85%) is consistent with the inferred ages of the points and indicates Marana and Haiwee occupations dating from ca. A.D. 600 to just before Euroamerican intrusions.

A striking display of site-use continuity is at INY-1534B with 50% of the hydration readings falling within the Marana interval and most of the remaining hydration rim values assignable to the previous Haiwee Period (Table 6.2). Notably, over 25% of the hydration measurements fall within the most recent half of the Haiwee era with only a few readings dating to earlier periods. The rock rings and intensive milling/seed processing activities exhibited at the Coso Junction Ranch site appear early in the Haiwee era, but their most intensive use dates mainly to late Haiwee times. The site shows strong continuity with and transition to the following Marana interval as exhibited by both the hydration rim profile and the frequency distribution of time-sensitive projectile point forms.

Similar settlements with rock rings are found throughout the Coso Range but have been documented most frequently in the areas of Fossil Falls north of Little Lake (Garfinkel 1976; Gilreath 1992), the Volcanic Tablelands in the northern Owens Valley (Basgall and Giambastiani 1995), and the western El Paso Mountains (Rogers 2004). In the latter area, hundreds of rock ring houses pepper the basalt benches of Black Mountain (Rogers 2004; Schiffman and Garfinkel 1981a: 3–28). Intensive study of these types of settlements has suggested that they served in some cases as threshing floors where seeds were processed using flash-burning methods (Basgall and Giambastiani 1995; Delacorte 1995). Direct flotation evidence indicates mass harvesting and threshing of rice grass (*Achnatherum hymenoides*), cattail (*Typha* spp.), goosefoot (*Chenopodium* spp.) and blazing star (*Mentzelia* spp.) seeds.

Basgall and Delacorte (2003) have recently attempted to clarify trends in late prehistoric plant use in eastern California. They suggest that a regionwide expansion of diet breadth and intensification of small seed resources involved a change in the technology used in the collection and processing of these resources. They argue that cutting and mass collecting of green, dryland hard seeds provided a considerably higher return than was possible using the former method of seed beating. They believe this pattern began in Haiwee times and substantially increased through the Marana era and into the Protohistoric period (Basgall and Delacorte 2003: 232–235).

Intensive Piñon Procurement: Archaeologists generally argue that adaptive shifts in technology facilitated intrusion into new environmental zones (Basgall 1993; Bettinger 1976, 1977, 1994). Sites located along the Sierra Crest and in the upland piñon forests of the Coso Range (Hildebrandt and Ruby 2000, 2003) are correlated with a new storage technology associated with intensive green-cone piñon pinenut procurement. These archaeological sites appear to represent an initial Numic presence, indicated by the introduction of these piñon storage features and Rose Spring and Desert Series points (Bettinger 1976; Garfinkel et al. 1980; Hildebrandt and Ruby 1999:30). These rock ring caches (piñon storage features) date to the onset of the Sawtooth/Haiwee era, with a continued emphasis and intensification culminating in a preponderance of such facilities during the Chimney/Marana Period (see discussion in Chapter 5).

Single-component sites testify that an emphasis on intensive green-cone piñon procurement activities occurred in the far southern Sierra, beginning ca. A.D. 600 (see Chapter 5). A compendium of all Coso obsidian hydration measurements for the Sierra Crest generally supports that notion, and exhibits continuity in rim frequencies with substantial representation of hydration rims associated with the Sawtooth (48%) and Chimney (24%) intervals (Table 6.2). Significant numbers of readings also date to the earlier Canebrake Period. These earlier rims are associated with other classes of loci interpreted as probable pre-Numic expressions emphasizing upland logistical hunting and apparently less intensive (brown-cone) piñon exploitation (see Chapter 5 and discussion below under Numic and Pre-Numic Contemporaneity).

Haiwee Period Intensive Small Game Procurement: The Bickel Site (KER-250) is located just above Last Chance Canyon on the western slopes of the El Paso Mountains, only about 15 miles from the Sierra Crest sites, at an elevation of 3100 feet amsl. The site is typical of a number of single-component, pure, Rose Spring (Haiwee Period) expressions found in eastern California (Delacorte 1994). The site was the subject of an extended study in which 15 one-by-two-meter units were excavated to a depth as great as 1.7 meters (McGuire et al. 1982). Most (11 of 14) projectile points were Rose Spring forms. Three radiocarbon dates fell within the Haiwee Period (A.D. 600 to 1300). The site was apparently a seasonally occupied fall encampment, whose occupants sought to procure large numbers of jackrabbits through communal hunts. Fully 73% of all identifiable faunal material (737 of 1011) and most (97%) of the less identifiable and more fragmentary material (4987 of 5161) were apparently of jackrabbits. Coso obsidian hydration rim distributions testify to a Haiwee occupation with a particularly intense expression during the recent half of this era (Table 6.2).

The Ash Creek Site (INY-3812) lies at the eastern foot of the Sierra just above the western rim of Owens Lake at the southern end of Owens Valley and just north of Ash Creek at an elevation of 4000 feet amsl (Gilreath and Holanda 2000). The site contained a 40- to 50-cm-thick midden, rich in fire-cracked rock and dietary faunal remains. Chronological data largely support an occupation restricted to the Haiwee interval. Of the 104 time-sensitive points recovered, 99 were of the Rose Spring Series. The remaining classifiable points included three Cottonwood and two "ears" from Humboldt Basal-notched bifaces. *Olivella*, *Mytilus* (mussel), and *Haliotis* (abalone) shell beads, when diagnostic, also supported temporal placement from ca. A.D. 600–1300. Radiocarbon dates provided a maximum age range from 1300 to 725 B.P. (midpoints of calibrated radiocarbon ages). Hydration rims on Coso obsidian correlate well with other indicators, suggesting that the most intense occupation occurred in late Haiwee times with a less intensive episode dating to the early Haiwee era (Table 6.2).

The faunal profile indicates a diet in which artiodactyls were an important component. Unusually large numbers of grebes and lagomorphs were also a significant element of the archaeofaunal collection. Grebes are most numerous in the area during the fall when they congregate in enormous numbers during their migration to feed on brine shrimp in saline lakes with brackish waters. A seasonal occupation associated with a water-impoverished environment is suggested for the Ash Creek site (Gilreath and Holanda 2000).

Toolstone Access, Exchange Relations, and Territoriality: A labor-intensive Numic adaptation ("processor strategy") is also consistent with the "adaptive pose" for Coso obsidian exploitation patterns (cf. Bettinger and Baumhoff 1982, 1983; Ericson and Glascock 2004). Obsidian quarry sites, known as pit or bench mines, are found on the lower and middle benches of West Sugarloaf Mountain in the Coso Range (Elston and Zeier 1984: 59, Figure 9; Garfinkel et al. 2004). These quarries date to an apparently transitional period, falling mainly within the late Haiwee era but also continuing through the Marana interval. High quality, more easily accessed lag sources of surface Coso obsidian by this time were perhaps either exhausted or being monopolized by competing pre-Numic populations (Eerkens and Rosenthal 2004). The exploitation of Coso obsidian during this brief period may have necessitated more labor-intensive methods than previously employed. Therefore, obsidian quarrying operations during the late Haiwee and Marana intervals resulted in the bench- and pit-mining operations initially identified by Elston and Zeier (1984). Recent research confirms that such an intensive episode of obsidian stone quarrying dates precisely to this brief time period (cf. Garfinkel et al. 2004). Hydration readings from Maggie's Site represent these pit mining activities (Table 6.2).

Style and Cultural Tradition: Rock art appears to have been of little interest to Numic populations, as attested by a notable absence in the ethnographic literature. Copious details are found covering many other, often more esoteric, subjects characteristic of the Great Basin Shoshoneans (Driver 1937; Steward 1938), with no real discussion of this subject (cf. Quinlan and Woody 2003). The in-migration by Numic peoples may be displayed in the Coso Range by the surprising abundance of simple scratched-style rock art designs found most prominently in the upland piñon forests of the Cosos (Bettinger and Baumhoff 1982; Gilreath 2003; Hildebrandt and Ruby 2003).

This style of rock art, known as Great Basin Scratched, is generally presumed to be associated with Numic peoples and has a suggested age ranging from 1000 to 500 B.P. (Heizer and Baumhoff 1962; Quinlan and Woody 2003; Nissen 1974, 1982; von Werlhof 1965). Scratched rock art in the Cosos, and in many other locations in the Great Basin, is often superimposed on or spatially associated with earlier pre-Numic art (Bettinger and Baumhoff 1982; Quinlan and Woody 2003). The defacement or embellishment of presumably pre-Numic rock art images is suggested to have been a means of "socializing" the landscape (Quinlan and Woody 2003). Scratched rock art was possibly employed to negate the perceived malevolent magic associated with the older Coso glyphs, to disrupt the hunting activities of the precursor or competing pre-Numic populations, and to secure the area for Numic use (Bettinger and Baumhoff 1982; Quinlan and Woody 2003; Steward 1933). As such, these simple engraved images may represent an effort by incoming Numic populations to mark the monuments of the preceding population. Corollary with the predominance of scratched style rock art in the upland piñon forests of the Coso Range is a dearth of typical Coso Representational Style rock art in the same area (see below).

It also appears that some 500 years after the end of the Coso Representational Style petroglyph tradition, a dramatic upsurge occurred in the production and design of elaborate painted art (Coso Painted Style), correlating with a period of cataclysmic cultural stress (see Chapter 5). During the historic era, Euroamerican disruptions of the aboriginal economy, epidemic disease, famine, and genocide all contributed to the inauguration of and resurgence in the most recent and late-dating manifestations of rock art. Coso Style paintings copied some of the earlier design elements of the Coso Representational Style petroglyphs (cf. Chapter 5), yet also incorporated new and novel concepts probably influenced by Euroamerican activities and also related to the Ghost Dance (Schiffman et al. 1982; Stoffle et al. 2000).

Summary of Numic Occupation Patterns

Similarities between Marana/Chimney Period and late Haiwee/Sawtooth expressions support a close linkage for the Coso Junction and Chapman sites. The use of open-air seed camps with house rings is also a trait common to the Haiwee and Marana eras (Basgall and Delacorte 2003; Byrd and Reddy 2004; Hillebrand 1972), thus suggesting a smooth transition and cultural continuity from one into the next.

Notably at the onset of the Haiwee Period, upland green-cone piñon camps were first established — an integral element of the following Marana-interval subsistence-settlement system in eastern California (Gilreath 2000; Hildebrandt and Ruby 2003). Archaeological sites along the Sierra crest and in the small island of Coso Range piñon forest attest to the timing and character of this pattern. Throughout eastern California, including the central Owens Valley, intensive piñon exploitation marks the beginning of the Haiwee Period (Bettinger 1976). Zeanah (2002:251), in an overview of prehistoric piñon use throughout the Desert West, recognizes this pattern as atypical of other areas of the Great Basin. In explaining this unusual situation he argues that "better (piñon) base camps locations were already occupied and … (piñon) collection opportunities constrained" because of increased population pressure. Such "demographic packing" would appear to have been a function of either *in-situ* population growth or in-migration. The latter condition appears to have been the most likely, given location and timing, and was probably a function of Numic population movements.

Evidence from the Bickel and Ash Creek sites, as well as from a number of similarly postured Haiwee Period sites (Delacorte 1994; Delacorte and McGuire 1993; Gilreath and Holanda 2000; Williams 2004), attest to the mass harvest of easily procured small game animals particularly lagomorphs and grebes. Both prey animals could be hunted cooperatively with large nets and procured in substantial numbers. Since they are naturally abundant and reproduce rapidly they would be especially susceptible to mass capture and sustained harvests (Delacorte and McGuire 1993:286). Specialized hunting camps indicate a rather distinctive adaptive pose represented only during this time span and notably different from earlier or later settlements. These hypothesized Numic immigrant base camps (*sensu* Delacorte 1994) appear in locations that were previously unoccupied. Bettinger (1994:48), in fact, suggests that the "Numic spread" was accomplished by just such small groups that pioneered unused or sparsely used, marginal areas in relatively out-of-the-way corners of pre-Numic territory.

Apparently these Numic sites were located in settings not inhabited earlier or later in time. Delacorte (1994) notes that such sites are situated in such a way as to provide access to a fairly limited range of targeted resources (such as grebes and lagomorphs). He further argues that an initial period of focused small game hunting largely preceded the intensive plant exploitation patterns of the in-migrating Numic peoples. Madsen (1986) largely agrees with this model of initial Numic subsistence patterns, and in fact offers a scenario where this intrusion was fostered by the Numic introduction of the bow and arrow (cf. Delacorte 1995).

He argues that the Numic migration and the introduction of the bow and arrow could have quickly led to a depletion of large game. Such depletion could come from increased predation from greater numbers of hunters (pre-Numic plus Numic populations) and from the hunters' increased effectiveness — using the bow and arrow over the former dart and atlatl technology. With reduction in large game populations, a corollary increase in the exploitation of small game would be expected. Madsen specifically suggests that a spike in rabbit hunting would occur, since these animals were normally hunted with drive techniques and their numbers would not be significantly affected by the introduction of the bow and arrow. Just such a pattern appears to be characteristic of sites in eastern California (cf. McGuire et al. 1982; Williams 2004).

That short-lived Numic pattern is thought to date to a time from 1400 to 1000 B.P., just before the expansion of Numic groups northward and eastward (Delacorte 1994; Garfinkel et al. 2005). It appears then that Numic occupation may have begun early in the Haiwee era but have become most intensive in late Haiwee times.

The Pre-Numic Pattern — *In-Situ* Cultural Development and Disruption

Antecedent to the Numic occupation are settlements characteristic of what seems to be a different cultural tradition. A variety of archaeological sites (see discussion below) apparently exhibit cultural materials thought to represent such occupations.

Little Lake Area

Little Lake is a springfed body of fresh water located within a small area known as Rose Valley at the southernmost end of the Owens Valley. Rose Valley lies between the eastern scarp of the Sierra Nevada on the west and the Coso Range on the east. Little Lake is approximately five kilometers from the abundant sources of Coso obsidian associated with Sugarloaf Mountain in the Coso Range within the confines of the China Lake Naval Air Weapons Station. The antiquity of this natural oasis has been examined by Mehringer and Sheppard (1978), and determined to be no less than 5,000 years in age. The Little Lake area is about 20 kilometers east of the Rockhouse Basin and Kennedy Meadows sites.

Stahl Site (INY-182): Studies at the Stahl site, less than a kilometer north of Little Lake (Harrington 1957; Meighan 1981; Pearson 1995; Schroth 1994), document intensive settlement during the Little Lake Period. Projectile point forms, hydration rim readings, and radiocarbon dates all testify to sustained occupation spanning the period from ca. 6500 to 1500 B.C. Hundreds of Pinto points, as well as a smaller number of Silver Lake and Lake Mojave types, were recovered during Harrington's early investigations. The overwhelming abundance of Pinto points testifies to that point type's use. More recent obsidian studies, initially conducted by Meighan (1981) and reevaluated by Pearson (1995), bolster confidence in this interpretation of the Stahl Site chronology.

As represented in Table 6.3, Stahl Site hydration rims are associated nearly exclusively with the Little Lake Period (7.6–16.0 microns), with only a few readings during the following Newberry Period. Twenty-nine Elko series points were recovered from the Stahl site indicating some, albeit more limited, Newberry Period occupation. No hydration rims were identified for the following Haiwee or Marana periods. Only four points were recovered dating to the latter periods — one for the Haiwee Period and three of Marana age. Hence, the Stahl Site itself seems to have been largely unoccupied during the late prehistoric eras (after A.D. 600).

Stahl Site Rockshelter (INY-205): The Stahl Rockshelter is a small cave in a lava outcrop adjacent to the northwest corner of the Stahl Site. Its obsidian hydration rims also fall mostly within the Little Lake Period (Table 6.3). Yet, a larger percentage of hydration rims than at the neighboring Stahl

Table 6.3 Distribution of Coso Obsidian Hydration Rims for Little Lake/Newberry and Early Haiwee Assemblages in Eastern California.

Lowland Coso Obsidian Hydration Rims*	<3.7	%	3.8–4.5	%	4.6–5.2	%	5.3–7.6	%	7.7–16.0	%	Totals
Period Designations	Marana		Late Haiwee		Early Haiwee		Newberry		Little Lake		
Stahl Site (Iny 182)	0	0	0	0	0	0	8	6.90	108	93.10	116
Stahl Site Cave (Iny 205)	2	8.00	0	0.00	0	0.00	4	16.00	19	76.00	25
Little Lake/*Pagunda* (Iny 3826)	8	19.05	1	2.38	1	2.38	28	66.67	4	9.52	42
Rose Spring (Iny 372, Locus 1)	8	4.57	14	8.00	16	9.14	71	40.57	66	37.71	175
Portuguese Bench (Iny 2284)	21	4.34	33	6.82	109	22.52	298	61.57	23	4.75	484
Coso Volcanic Field	80	2.04	118	3.01	617	15.75	1392	35.53	1711	43.67	3918
Single Component Coso Petroglyphs	6	6.82	12	13.64	21	23.86	19	21.59	30	34.09	88
Lubkin Creek (Structures 11, 12, 14)	4	8.70	6	13.04	13	28.26	15	32.61	7	15.22	46
Wide Humboldt Basal-notched Bifaces	1	2.94	1	2.94	7	20.59	24	70.59	1	2.94	34

NOTE: Readings for the Little Lake area are taken from Meighan (1981) and Pearson (1995). Rose Spring, Locus 1 readings are from Yohe (1992) and include 20 new readings on Rose Spring points analyzed for this study. Coso Petroglyph data are from Gilreath (1999) and Garfinkel (2003). Additional data are from personal communication with Alexander "Sandy" Rogers (2004). Lubkin Creek data are from Basgall and McGuire (1988). Portuguese Bench site hydration readings include smallest of multiple hydration readings on single specimens, readings identified as outliers included. Data sources are Allen (1986) and Whitley (1988). Coso Volcanic Field data are from Gilreath and Hildebrandt (1997). Rims greater than 16.0 microns are not included in this summary. Humboldt Basal-notched biface data are from Garfinkel and Yohe (2004). Percentage distributions for particular chronological periods in excess of 20.00% are underlined in bold to emphasize substantial activity as represented by hydration rim frequencies.

*Readings included are for sites with hydration rim measurements on Coso obsidian situated at elevations from 3000 to 5000 feet amsl and using the lowland Coso obsidian hydration rate parameters suggested in Chapter 5 of this book.

Site also reflect occupation during the following Newberry Period. The Stahl Rockshelter was largely abandoned throughout the Haiwee Period, but limited occupation returned with the Marana Period. Occupation during Marana times is represented by two hydration rims, two Desert Side-notched and four Cottonwood Triangular points, Owens Valley Brownware ceramic sherds, European blue and red glass beads, and simple scratched and pecked drawings of bighorn sheep and native hunters (Grant et al 1968:94; Harrington 1957; Pearson 1995). The latter glyphs are thought to be more akin to Coso Style paintings both in style and dating than to the older Coso Representational petroglyphs (Austin 2005; Garfinkel et al. 2005).

***Pagunda* (INY-3826):** On the edge of Little Lake, Pearson (1995) identified an archaeological site believed to represent the Panamint Shoshone village site identified as *Pagunda* by Julian Steward (1938). The site manifests a paucity of Little Lake Period remains. Most of the occupation at *Pagunda* falls within the Newberry and early Haiwee periods evidenced by a large number of Elko-age hydration rims, some early Haiwee-age readings, and 26 Rose Spring series points (Table 6.3). Eight Desert Series points indicate occupation in the Marana Period. Again, curiously few hydration rims of the late Haiwee Period were identified at *Pagunda* (Pearson 1995: Figure 7; Byrd and Reddy 2004: Table 11).

Recent excavations at *Pagunda* corroborate these chronological interpretations (Byrd and Reddy 2004). Of 17 additional hydration rim readings on flakes, most (10) fall within the Newberry Period and five of the remaining rims are tightly restricted to the early Haiwee Period. Further bolstering these chronological interpretations are seven radiocarbon dates (Byrd and Reddy 2004:Table 14.3). All of these dates, when calibrated, fall within the early Haiwee era (A.D. 600–950).

Summary: The Little Lake area, in general, shows its most intensive and substantial cultural expressions during the Little Lake Period when Pinto points were most popular. Some Newberry era occupation is evident, especially at the *Pagunda* and Stahl Cave sites. The area appears to have been largely abandoned during the late Haiwee interval as attested by a marked decline in Coso obsidian hydration rim readings, radiocarbon dates, and time-marker artifacts. This hiatus comes to an end with a late prehistoric reoccupation of the area reflected by a number of hydration rims, pottery, Desert-Series points, historic glass trade beads and late dating rock drawings. Such a pattern is consistent with two distinct cultural traditions and an occupation punctuated by a period of diminished cultural activity or abandonment followed by an apparent population replacement.

Rose Valley Area

Just north of Little Lake, in the Rose Valley area, lie the Rose Spring (INY-372) and Portuguese Bench (INY-2284) sites. Rose Spring is located at the eastern base of the Sierra Nevada at the northernmost end of Rose Valley, just south of Haiwee Reservoir. On the other side of Rose valley, 12 kilometers south of Rose Spring and 13 kilometers east of the Rockhouse Basin and Kennedy Meadows sites, is Portuguese Bench (Figures 1.2 and 1.10).

Rose Spring (INY-372, Locus 1): The Rose Spring Site is one of most significant archaeological sites in eastern California (Yohe 1992). The site has played an important role in the development of western Great Basin culture history, since it was a rare example of a deep (3.7 m), open-air, site with an artifact-rich deposit that was both physically and culturally stratified. Seventeen radiocarbon dates are available for Locus 1 of the Rose Spring Site: five based on samples collected by Riddell in 1956 and analyzed by Clewlow et al. (1970); and 12 obtained by Yohe (1992), 10 dates fall within the Little Lake and Newberry periods, and seven within the Marana Period. Excavation levels dating to the Haiwee Period

(60–120 cm.) are bracketed by radiocarbon dates closely synchronous with the generally accepted end dates of this period. Yet radiocarbon assays, falling within the entire span of the Haiwee Period, are curiously lacking (cf. Byrd and Reddy 2004: 308–310; Yohe 1992:140).

Significantly, cultural activity at Locus 1 appears to take a dramatic turn during the Haiwee era. Seventy-eight percent of all obsidian hydration measurements (Table 6.3) are outside of the Haiwee Period and lie within either the Newberry or Little Lake periods (cf. Gilreath and Hildebrandt 1997:165–166, Figure 24). Those hydration readings suggest occupation dating between 5500 and 1500 B.P. Gilreath and Hildebrandt (1997:166) suggest that Locus 1 of Rose Spring was "foremost a late Newberry obsidian reduction workshop." Their analysis led them to conclude that, "most of the debitage in the site was generated during the Newberry Period occupations."

Eighty-three Rose Spring points, the site's most common point form, were recovered from Locus 1. These points confirm that the site was occupied during the Haiwee Period. Yet the archaeological record of the Haiwee Period at Locus 1 also reveals a decline in hunting and other cultural activity. Surprisingly, faunal data from Rose Spring demonstrated a sharp decline in hunting when the bow and arrow were introduced. The highest frequency of dietary faunal remains is from excavation levels *preceding* this dramatic decline. Specifically, bones of large ungulates predominate by number and weight in the levels dating to the late Newberry Period (Yohe 1992:140, Table 5; Yohe and Sutton 1999, 2000) and large mammal dominance resumed only in the Marana Period, when hunting almost reached the intensity attained previously.

Occupation during the Haiwee Period apparently shifted away from faunal exploitation, and although the site was certainly far from abandoned, activities represented at the Rose Spring site must have had a very different cast from any other time throughout prehistory (Yohe and Sutton 1999, 2000). Reduced site activity is also reflected in lowered amounts of flaked-stone debris in excavated levels dating to the Haiwee Period (Yohe 1992:230, Figure 45). Those levels see a significant decline in the number of obsidian flakes, decreasing as much as a third to a half of that retrieved from earlier Newberry Period occupations.

Excavation Unit G-1 drops from a peak of nearly 4,000 flakes per level during the Newberry Period to 1,000 flakes per level at the end of the Haiwee era. Unit X-3, although not containing recognized strata of the Newberry Period, produced debitage dating to the Haiwee Period and exhibited a steep decline in unit/level debitage volumes within that 600-year period. The unit produced a peak density of flakes per level (about 2,500 per level) at the beginning of the Haiwee Period and then quickly declined to a low of 1000 flakes per level at its end. Again, after decreased levels of activity during the Haiwee Period, flaked-stone densities in some excavation units returns almost to the quantities characteristic of the pre-Haiwee Period, with 2,500 to 3,500 flakes per level (Yohe 1992: Figure 45).

Using time-adjusted flaked-stone quantities for the Marana, Haiwee and Newberry periods would serve only to magnify the differences reported. Less debitage is normally expected in Marana-age occupations, since during this period smaller arrow points were being produced, requiring far less stone than did Newberry-age dart point production. Considering those factors, the amount of debitage recognized in Marana levels and the discrepancy with the preceding Haiwee deposit may be all the more significant. Site stratigraphy also corroborates the pattern of reduced cultural activity dating to the Haiwee Period with lighter-colored soil, less fire-affected rock, and fewer features (Yohe 1992).

Portuguese Bench (INY-2284): Portuguese Bench, situated at the foot of the eastern scarp of the Sierra Nevada, is an extensive village site with numerous structural remains and a large volume of flaked obsidian (Allen 1986; Whitley 1988). Excavation results have not been fully documented in a comprehensive site report. The nearly 500 obsidian hydration readings reported (Allen 1986; Whitley 1988) are similar to those of Locus 1 of the Rose Spring Site; most hydration measurements again fall within the Newberry Period, with substantial, yet declining, activity represented during the early Haiwee era (Table 6.3). This decline suggests that aboriginal activity continued to diminish during the latter portion of the Haiwee Period. The site was all but abandoned during Marana times.

Summary: Two large residential sites, located in Rose Valley, had their greatest occupation during the Newberry Period. Both Rose Spring and Portuguese Bench produced similar hydration rim distributions. Although Rose Spring contains significantly more occupational indicators from the Little Lake Period, the archaeological records at both sites suggest a major decline and occupational shift in the Haiwee Period. Occupation declined in the second half of the Haiwee Period at both localities. During the most recent Marana Period, Portuguese Bench saw minimal aboriginal activity. After this period of diminished activity at Rose Spring, considerable occupational debris and an expanded faunal inventory marks the onset of the Marana Period.

Coso Range Area

Lying mostly within the boundaries of the China Lake Naval Air Weapons Station, the Coso Range includes sites located within the Coso Volcanic Field and the petroglyph concentrations are largely restricted to the basalt flows within the northern base.

Coso Volcanic Field: Excavation results from 34 sites, coupled with hydration rim measurements from about 4000 obsidian artifacts, help clarify land-use patterns associated with obsidian quarrying and stone tool production within the Coso Volcanic Field. These data reveal an extensive use of the area through the last 10,000 years (Gilreath and Hildebrandt 1997). Yet peak use is restricted and dates only to the Little Lake and Newberry era, with a significant drop and concomitant land-use change documented to the Haiwee and Marana periods.

Occupation and site use in the Newberry, Little Lake, and earlier periods were focused chiefly on obsidian quarrying, artiodactyl hunting, and rock art production. The Little Lake Period saw short-term use of lag quarry deposits. Newberry Period activities shifted focus to off-quarry biface production and also expanded production work at primary outcrops of the highest quality obsidian. Such activities were apparently associated with a network of trans-Sierran obsidian exchange. Quarrying of obsidian dropped precipitously during the Haiwee Period in the Coso Volcanic Field itself (Gilreath and Hildebrandt 1997). Yet secondary reduction is amply documented for the early half of the Haiwee Period at the Portuguese Bench Site (see discussion above and Table 6.3). By the Marana Period evidence for local flaked-stone reduction is negligible and direct quarrying at the Coso obsidian sources was almost completely absent. During the Marana Period, the area became a nearly exclusive locus for intensive seed processing and plant collecting activities (Gilreath and Hildebrandt 1997).

Coso Range Petroglyphs: Hydration rim readings associated with single-component Coso Petroglyph sites have been obtained for 11 archaeological sites (Garfinkel 2003; Gilreath 1999). The total sample of 88 hydration rim readings documents a production chronology similar to other archaeological expressions dating principally to the Little Lake, Newberry, and early Haiwee periods (Table 6.3). As best as can be reconstructed, substantial production of Coso rock art occurred in the Little Lake Period and continued unabated through the Newberry Period. Activity in the Haiwee Period increased dramatically, but abruptly declined during the latter half of that time span and was virtually absent in the following Marana Period.

Summary: Intensive aboriginal activities, as documented in the Coso Range, clearly emphasized obsidian acquisition, ungulate hunting, and rock art production. Hydration rim frequencies appear to show that those activities were predominantly of early Haiwee, Newberry, and Little Lake age. A land-use shift occurred in the late Haiwee and Marana eras with a noticeable emphasis on the procurement of hard seeds and other key economic plant foods instead of artiodactyl hunting, the manufacture of associated weaponry, and the production of rock art (Table 6.3).

Southern Owens Valley Area

One of the best-documented Newberry-age residential settlements is Lubkin Creek (INY-30) in the southern Owens Valley (Basgall and McGuire 1988), where several well-built houses and their associated remains provide a clear picture of occupation during the Newberry Period. As has often been remarked, these remains include an emphasis on cached and curated articles (including bifaces, bone tools, and milling equipment), lending credence to the premise that sites of this period were seasonally reoccupied. Obsidian tool/debitage sources represented at Lubkin Creek, and other sites of similar age, indicate a wideranging and extremely expansive annual settlement round. From food remains (faunal material and plant macrofossils) the inference may be made that logistical forays were made to long-distance upland settings to procure resources (piñon nuts, large game) that were brought back to the base camp. Animal remains show an emphasis on ungulates, and well-made milling equipment documents the increasing importance of plant foods. Coso hydration rims from three structures of Newberry age at INY-30 show continued site use dating to the Newberry and early Haiwee periods (Table 6.3).

Style and Cultural Traditions: It has been recognized that most time-sensitive Great Basin projectile point types have popularity curves displaying a bell shape, which shows that most forms are usually replaced gradually by temporally overlapping forms. Such a gradual decrease, as evidenced by diminishing obsidian hydration frequencies and overlapping hydration reading distributions, has been noted as characteristic of most time-diagnostic southwestern Great Basin point series with the notable exception of the Wide Humboldt Basal-notched type (cf. Garfinkel and Yohe 2004; Jackson 1984).

The Wide Humboldt Basal-notched form appears to have been discontinued rather abruptly at ca. A.D. 800–1000 with lowland Coso hydration rim readings no smaller than 4.7 microns (Garfinkel and Yohe 2004: Figure 2, Table 1), placing that termination date within the span of the early half of the Haiwee Period. The Wide Humboldt Basal-notched type has a notable floruit from ca. 500 B.C. to about A.D. 1000 and has lowland hydration rims (on Coso obsidian) no larger than 7.7 microns (Tables 6.4 –6.6).

Such a brief, discrete, and marked chronological distribution of the form can be recognized by its tightly restricted distribution of hydration rims clustered about the mean (Tables 6.4 and 6.5). As a measure of such distribution, Wide Humboldt Basal-notched bifaces have the narrowest range (a spread of only 3.0 microns) of hydration readings of any chronologically diagnostic Great Basin point form (Table 6.6). The form also has one of the smallest coefficients of variation (0.13) (Garfinkel and Yohe 2004; Gilreath and Hildebrandt 1997). These bifaces are found in large numbers in caches (Basgall and McGuire 1988), as burial accoutrements (see below), and associated with game intercept drive sites (Garfinkel and Yohe 2004 and references therein).

Such unique associations support the notion that the form may have had special emblematic cultural, ritual, and/or religious significance. Since this form spans the Newberry and early Haiwee periods, it is precisely synchronous with the time when many prehistorians believe that pre-Numic populations were most active in the southwestern Great Basin (Delacorte 1994; Garfinkel 2003; Gilreath 1999). This form is also contemporaneous with an abundance of Coso Representational Style petroglyphs; the hunting of large artiodactyls using communal drive techniques; and extensive, logistical mobility patterns (Basgall and McGuire 1988; Gilreath 1999; Gilreath and Holanda 2000). The distinctive and rather abrupt termination of wide Humboldt Basal-notched bifaces may relate to cultural, technological, or sociopolitical factors affecting eastern California during this period. The seemingly abrupt discontinuation of the Wide Humboldt Basal-notched form, during its full fluorescence may also reflect a population replacement in the study area (Table 6.4). Given these considerations, the Wide Humboldt Basal-notched bifaces might be hypothesized as an ethnic signature for the pre-Numic people in eastern California.

Burial Patterns: Two recent compilations of prehistoric burial patterns in eastern California include the areas of the Owens Valley, Rose Valley, Death Valley, and Coso Range (Gilreath 2000; Gilreath and Holanda 2000:122–126, Appendix I). A total of 43 separate burials representing 48 individuals are recorded at 22 sites. The majority of these burials are primary interments; four are cremations. All the inhumations were semi- or loosely flexed.

In Owens Valley and Death Valley, the individuals buried tend to be placed on their right side; burials in the Coso Range were interred on their left. Cairns overlaid 15 burials; multiple bead lots occurred in three. Only in the late Newberry and early Haiwee periods is there a tendency for a rich array of grave goods (Table 6.7). Multiple projectile points, assorted artifacts, and, in one spectacular instance an extraordinary collection of 1,000 *Haliotis* wide-ring beads (placed in a shingled arrangement on the chest of the deceased child) are noted as grave offerings. Therefore, burials dating to the late Newberry/early

Table 6.4 Lowland Coso Obsidian Hydration Readings on Wide Humboldt Basal-notched Bifaces

Provenience	Obsidian Source	Number	Readings* (in microns)	Mean	Range
Rose Spring Site	**Coso	13	4.8, 4.8, 5.7, 5.7, 5.9, 6.0, 6.0 6.2, 6.5, 6.6, 6.7, 7.0, 7.6	6.1	4.8–7.6
Coso Volcanic Field	#Coso	8	(2.3), 4.8, 5.7, 5.7, 5.9, 6.0, 6.8, 7.5, 7.7	6.3	4.8–7.7
Lubkin Creek Site	+Coso	11	4.7, 4.7, 4.9, 4.9, 5.3, 5.5, 5.8, 5.9, 6.0, 6.1, 6.9, (8.2)	5.5	4.7–6.9

KEY: *Readings in parentheses have not been included in the statistics for mean and range.
Not chemically sourced; inferred from location, and +sources determined using x-ray fluorescence.
** Not chemically sourced; inferred from location.

Table 6.5 Statistical Summary of Obsidian Hydration Data on Wide Humboldt Basal-notched Bifaces.

Statistical Summary	Lubkin	CVF	RS
Number of cases	11	8	13
Mean (in microns)	5.5	6.3	6.1
Standard deviation	1.0	1.0	0.8
Coefficient of variation	0.18	0.13	0.13

KEY: Lubkin=Iny 30, CVF=Coso Volcanic Field, RS=Iny 372, Rose Spring, Locus 1. Outliers excluded.

Table 6.6 Comparative Obsidian Hydration Ranges (in microns) for Projectile Points from the Coso Region.

Projectile Point Type	Number	Micron range	Spread	Mean	SD	CV
Desert Series	12	1.2–4.7	3.5	3	1.2	0.40
Rosegate	20	3.6–6.9	3.3	5.2	0.8	0.15
"Wide" Humboldt Basal-notched Bifaces	33	4.7–7.7	3.0	6.3	1.0	0.13
Thin Elko	12	6.0–9.3	3.3	7.4	1.0	0.13
Thick Elko	8	8.7–18.9	10.2	12.3	3.3	0.27
Pinto	12	9.1–21.5	12.4	14.2	4.3	0.30
Great Basin Stemmed	21	8.7–17.8	9.1	12.9	2.7	0.21
Concave Base	2	13.4–21.1	7.7	17.3	5.4	0.31

KEY: Humboldt Basal-notched Biface measurements abstracted from Garfinkel and Yohe (2005). Other point type data from Gilreath and Hildebrandt (1997).

Haiwee periods are different from those of other chronological intervals, yet they are similar as a class unto themselves (cf. Gilreath and Holanda 2000). It is tempting to suggest that this similarity indicates a distinctive cultural expression. However, most burial descriptions are brief and incomplete, weakening the validity of such a premise.

Nevertheless, prior researchers have identified a ritual association for projectile points dating to the Late Newberry and Early Haiwee periods as represented in the petroglyphs in the Coso Range (Garfinkel and Pringle 2004; Hildebrandt and McGuire 2004). Burial associations with multiple Wide Humboldt Basal-notched bifaces (Garfinkel and Yohe 2004), and frequent Elko and Rose Spring points (Garfinkel and Pringle 2004) may represent a preoccupation with hunting weaponry dating to this time span. Notably, a technological analysis of five Humboldt Basal-notched bifaces recovered from a burial at the Barnett Site, Nye County, Nevada, in the Amargosa Desert, led researchers to conclude that these implements were manufactured specifically for inclusion with the burial (Muto et al. 1976:275). Additionally, Garfinkel and Pringle (2004) have documented the depiction of Rose Spring and Eastgate points within panels of Coso Representational drawings thought to be contemporaneous with their peak production interval during the Haiwee Period (A.D. 600–1300).

Summary

Time-sensitive projectile points, obsidian hydration rims, and radiocarbon dates document substantial, long-term occupations in the areas of Little Lake (Stahl, Stahl Cave, and *Pagunda*), Rose Valley [Rose Spring (INY-372), Portuguese Bench (INY-2284)], the Coso Range, and Lubkin Creek (INY-30) (Allen 1986; Basgall and McGuire 1988; Byrd and Reddy 2004; Hildebrandt and Gilreath 1997; Lanning 1963; Pearson 1995; Schroth 1994; Yohe 1992). What is striking about the dating of these sites is that there is a common pattern indicating an evident disjunction in the midst of the Haiwee Period (cf. Byrd and Reddy 2004:309; Pearson 1995:110, Figure 7). By paraphrasing the insights of Gilreath and Holanda (2000) and relying on some of the suggestions of Delacorte (1994, 1995), eastern California prehistorians have concluded that the Haiwee Period is characterized by substantial cultural changes, yet also provides significant evidence of continuity and stability.

Such patterns could equate with a population turnover in the midst of the Haiwee Period. Earlier pre-Numic occupations show continuity from the Newberry Period into the early Haiwee, and intruding Numic groups may manifest their first presence ca. A.D. 600 with the inception of the Haiwee Period and Rose Spring points (cf. Delacorte 1995). In the early portion of the Haiwee Period, many

Table 6.7 Late Newberry/Early Haiwee Age Burials From Eastern California.

Site	Age	Sex	Position	Assoc. Points
Tibbi Opo (Iny 4646)	Adult	?	Semi-flexed	2 or 3 HBNs
Lubkin Creek (Iny 30)	Adult	M	Cremation	8 HBNs cache???
Manzanar (Iny 4864/H)	Y. Adult	M	Tightly flexed	2 E and 1 RS
Pa Doya (Iny 1991)	Y. Adult	F	Not reported	4 HBNs
Rose Spring (Iny 372)				
Burial 2	Adult	?	Semi-flexed	2 E, 4 RS, 2 HBN
Burial 4	Child	?	Tightly flexed	1 E
"Grants Tomb" (Iny 2847)	2 Adults	M & F	Flexed	2 E
Barnett (Nye County, Nevada)				
3 Burials	?	?	Loosely Flexed Flexed	5 HBNs

NOTES: *Tibbi Opo* (Iny 4646) is the subject of a monograph by Burke et al. (1995). Lubkin Creek is reported by Basgall and McGuire (1988). Burton (1996) discusses the results of prehistoric investigations at Manzanar. *Pa Doya* is covered in the study by Markos et al. (1995). The Rose Spring site is most thoroughly discussed in Lanning (1963) and Yohe (1992). Clewlow et al. (1995) tested "Grant's Tomb" (Iny-2847). The Barnett site is reported by Muto et al. (1976).

settlements show an extension from the preceding late Newberry Period practices. Substantial Haiwee Period habitation sites, located atop Newberry-age deposits (e.g., Rose Spring, Portuguese Bench, *Pagunda*), indicate site-use continuity. Chronological associations of burials found in the bases of these Haiwee middens are also curiously ambiguous as to whether they best relate to late Newberry or early Haiwee affiliations (e.g., *Tibbi Opo* [Burke et al. 1995], Manzanar [Burton 1996], and *Pa Doya* [Markos et al. 1995]) (Gilreath and Holanda 2000:77). Wide Humboldt Basal-notched bifaces are frequently found with both Elko and Rose Spring series forms (cf. Burton 1996). At the Lubkin Creek Site, recognized as one of the most thoroughly investigated late Newberry occupation sites, Elko series points were recovered in numbers equivalent to either Humboldt Basal-notched or Rose Spring forms (Basgall and McGuire 1988:130). Hence, late Newberry-age and early Haiwee-age sites are often largely indistinguishable.

It can be postulated that pre-Numic sites were largely Newberry, earlier in age, and show site-use continuity through the early Haiwee interval. Pre-Numic sites apparently have late Haiwee manifestations that evince declining levels of activity, decreasing use, and in many cases, an abrupt termination of cultural activities (cf. Delacorte 1994).

Numic Continuity or Population Replacement?

A number of prehistorians argue that the Numic people were longtime occupants of eastern California and the southwest Great Basin, in part since linguists point to this general area as the original Numic homeland. Yet if such a reconstruction were accurate, the archaeological record should reveal substantial evidence for a relatively continuous and unbroken cultural record, with gradual, in-place, changes, and marked continuity. The latter would be similar to the patterns identified for the archaeological assemblages noted in the former homeland of the Tubatulabal. As mentioned in Chapter 1, rock art data have been central in arguments positing a continuous ethnic thread. Warren (1984:384) states this perspective rather clearly in stating that

> ...the Coso petroglyphs represent a cultural tradition that persists through several archaeological phases, suggesting that there is considerably more cultural continuity than recognized in the archaeological assemblages. The artistic tradition and presumed ceremonial tradition represented by the Coso petroglyphs suggest that the historic Shoshonean peoples of the Coso Mountain area have cultural origins that extend far back into the local prehistoric sequences.

Such an interpretation, if accurate, would support in-place Numic development and cultural continuum (cf. Garfinkel 1982; Grant et al. 1968; Pearson 2003; Whitley 1994; 1998:53–60). Hence, a review of the rock art record in the Coso region is central to this discussion.

Coso Representational Petroglyphs and the Numic Intrusion

Rock art is potentially one of the most sensitive indicators of ethnic affiliation and cultural identity. Subtle changes in religious rituals are frequently referenced as manifestations of dramatically disparate ethnic identifications (Barth 1969). Many prehistorians have argued that it may be more difficult to recognize prehistoric population movements in the southwestern Great Basin than elsewhere, since hunter-gatherer economies in this area continued in place. In other areas of the Great Basin, agriculturists replaced foragers and exhibited more striking differences in material culture (cf. Madsen 1994).

Southwestern Great Basin rock art sites vary in subject matter, style, spatial distribution, and archaeological context and, therefore, may be some of the most informative and persuasive forms of archaeological evidence. Because rock art captures and reflects human interactions with natural and cultural environments, it enlightens its viewers as to aboriginal population movements and cultural identities. The rock art tradition of the Coso Range is distinctive, if for nothing else, due to its sheer abundance and surprising degree of realism.

Conservative projections, based on selective sample surveys, indicate an excess of 100,000 individual elements concentrated in an area of less than 90 square miles (Gilreath 1999; Hildebrandt and McGuire 2002). The Coso Range therefore contains one of the greatest concentrations of petroglyphs in all of North America (Grant et al.1968). Between 60 and 90 percent of these representations are realistic portrayals of the quarry, technology, and ritual paraphernalia associated with hunting bighorn sheep. Bighorn depictions are commonly found throughout the Desert West, yet the number of images within the Cosos is thought to surpass the total number of sheep drawings for all other regions combined (Grant et al. 1968:34). Grant et al. (1968:115) comment that "What is so astonishing about the Coso Range rock art complex is that it apparently developed in almost complete isolation, an island of specialized art tradition."

Coso Petroglyphs: Function, Authorship, and Dating

The functional interpretation of rock art in the Great Basin has been an especially contentious issue (Hildebrandt and McGuire 2002; Rector 1985; Whitley 1998). Still, consistent correlation of Great

Basin rock art sites with game trails, ambush locations, dummy hunters, hunting blinds, game corrals, and the depiction of artiodactyls (overwhelmingly mountain sheep) and hunting weapons argues strongly for its use as sympathetic magic to help ensure successful hunting of big game (Heizer and Baumhoff 1962; Hildebrandt and McGuire 2002; Grant et al. 1968; Nissen 1975, 1982, 1995; T. Thomas 1976; von Werlhof 1965). Additionally, although the functional interpretation of Great Basin rock art has been contentious, functional ascription may be largely irrelevant to the present purpose since age, style, motif, association, and other traits should suffice to define ethnic associations.

Despite the ubiquity of rock art in many areas of the Great Basin and even more so in the Coso Range, Native Americans in recent eras denied authorship of the rock art and any knowledge of its meaning (Heizer and Baumhoff 1962; Steward 1929, 1968). Great Basin ethnography contains little information concerning rock art, suggesting that much of it is probably pre-Numic. Julian Steward (1968:viii–x) went so far as to assert that the Coso petroglyphs are one of the more striking examples of either cultural loss or the replacement of one cultural group by another in the general study area. He states that

> ...the Shoshonean Indians...knew nothing of the authorship or meaning of the petroglyphs, and their culture seemed unlikely to manifest itself in this medium. ...none of the various Shoshonean activities or categories of culture, except basketry, were expressed in any art form. ...Shoshonean subsistence activities have no apparent relationship to this art. ...this rock art signifies cultural loss in the area, owing either to deculturation of the present inhabitants or to *the earlier presence of a different people* [italics added].

Grant et al. (1968) were the first researchers to attempt to date the Coso petroglyphs (Figures 6.4–6.6). Based on superposition of images, degree of patination, and seriation of hunting weaponry and other subject matter, Grant et al. developed a temporal ordering indicating that most of this rock art was produced between 3000 and 1000 B.P. (ca. 1000 B.C. to A.D. 1000). They posit that during the Early Period, imagery was chiefly abstract with simple renderings of sheep and highly stylized atlatls. The Transitional Period included many depictions of bighorn sheep and shield designs, "medicine bags" (most likely bighorn sheep hunters in disguises; see Heizer and Hester 1974; Nissen 1982), dogs attacking sheep, simple anthropomorphs, and a combination of dart-and-atlatl and bow-and-arrow images. Late Period renderings were the most realistic, and indeed the largest and most complex designs, with bighorn sheep drawn in large scale with great care.

Late Period petroglyphs of bighorn sheep contained the uniquely characteristic navicular bodies, flat backs, and full front-facing, head-on horns, often with hooves and ears added. Grant et al. hypothesized that the elaborate anthropomorphs and depictions of weaponry (bows and arrows) continued until ca. A.D. 1000. The last period of petroglyph production is represented by a simple form of rock scratching that defaced or embellished the earlier drawings (Bettinger and Baumhoff 1982; Coombs and Greenwood 1982; Quinlan and Woody 2003). That style has been termed Great Basin Scratched (Heizer and Baumhoff 1962; Nissen 1974; von Werlhof 1965).

Gilreath (1999) recently used obsidian hydration readings associated with 43 Coso Range sites to evaluate the various dating schemes for the Coso petroglyphs (cf. Garfinkel 2003). Her research points to the Haiwee Period (A.D. 600–1300) in the local chronological sequence as the time span when the greatest number of rock art sites were produced. These Haiwee Period sites contain chiefly representational motifs (65%). Earlier sites are dominated by abstract designs. Gilreath's study identified a rather abrupt decline and termination for the petroglyph drawings dating to no later than A.D. 1300 (with 94% of the 505 obsidian hydration rim readings in her study falling into earlier time spans). Her work also indicates that Coso rock art is almost exclusively a pre-Marana Period (A.D. 1300–1850) expression (greater than 3.7 microns of hydration on Coso obsidian), with a distinctive Haiwee Period emphasis (A.D. 600–1300, or 3.7–4.9 microns). Single-component Coso petroglyph sites appeared in the Mojave Period dating ca. 8000 B.C. (based on mean hydration rims).

Further Evidence for Dating Coso-Style Petroglyphs

Recent independent testing of Gilreath's dating scheme supported its general validity (Garfinkel 2003). Evaluation of the archaeological associations of stylistically similar drawings just outside the Coso Range provided temporally equivalent Coso obsidian hydration readings. Further validation of the dating scheme comes from an analysis of the projectile point drawings depicted in Coso petroglyphs. The drawings of realistically rendered points were interpreted as analogs of either Rose Spring Corner-notched or Eastgate Expanding Stem forms (Garfinkel and Pringle 2004). Garfinkel and Pringle (2004) argue that such depictions date the period of greatest rock art production to the Haiwee Period (A.D. 600–1300).

Independent assessment of the temporal span for Coso rock art also comes from their archaeological context. Excavation of an aboriginal settlement (Site 14-5488), physically associated with the largest concentration of Coso Representational Style glyphs, in Renegade Canyon (Gilreath 2000:51–61), produced temporally diagnostic projectile points overwhelmingly indicative of *only* the Newberry and Haiwee periods (3500–650 B.P.), with 16 Elko and 27 Rose Spring/Eastgate point forms. Manifestations of more recent occupation were especially meager — Marana Period materials were limited to only two Desert Side-notched points, a dozen sherds of pottery, and four beads. Therefore, the weight of evidence strongly indicates that the period of greatest abundance and production of Coso Representational rock art dates exclusively to the Newberry and Haiwee periods and abruptly terminates during the waning years of the latter interval.

Implications of Discontinuity

A long tradition of rock art production exists within the Coso Range; and this tradition appears to have begun more than 10,000 years ago based on the associated Coso obsidian hydration rims (Gilreath 1999). However, glyphs older than about 5,000 years ago appear to be dominated by abstract images (Figure 6.1). The majority of the later drawings date to a relatively brief fluorescence in the late Newberry and early Haiwee periods. During those eras, rock art may have played a much more dominant role in the lives of the Coso hunters than perhaps elsewhere in the western Great Basin (Garfinkel 2003; Garfinkel and Pringle 2004; Gilreath 1999:15). The distinctive style of bighorn depictions and the degree of realistic representation are unique characteristics of Coso rock art and an unusual feature of the images in the region. No other area of the Great Basin displays such emphatically realistic imagery, nor are bighorn sheep depictions rendered elsewhere in this unusual style (Grant et al. 1968; Schaafsma 1986). Grant et al. (1968:16–17) emphasize that "The drawings in this country cover a very long time span, and for the whole period the art tradition remained remarkably stable. We are certain that they are the work of the same people." This "self-contained tradition of unique style and location has such an extreme degree of thematic consistency, it must have represented a unified belief system with a standardized iconographic vocabulary" (Turpin 1990). Therefore, this complex and rather sophisticated art style can best be seen as an autochthonous development.

Abandonment of rock art signals significant changes in ritual practice and cultural loss (cf. Steward 1968). The abrupt discontinuation of the Coso petroglyph tradition while it was in full fluorescence and its replacement by a simple scratched rock art style could signal disruption by an exotic population

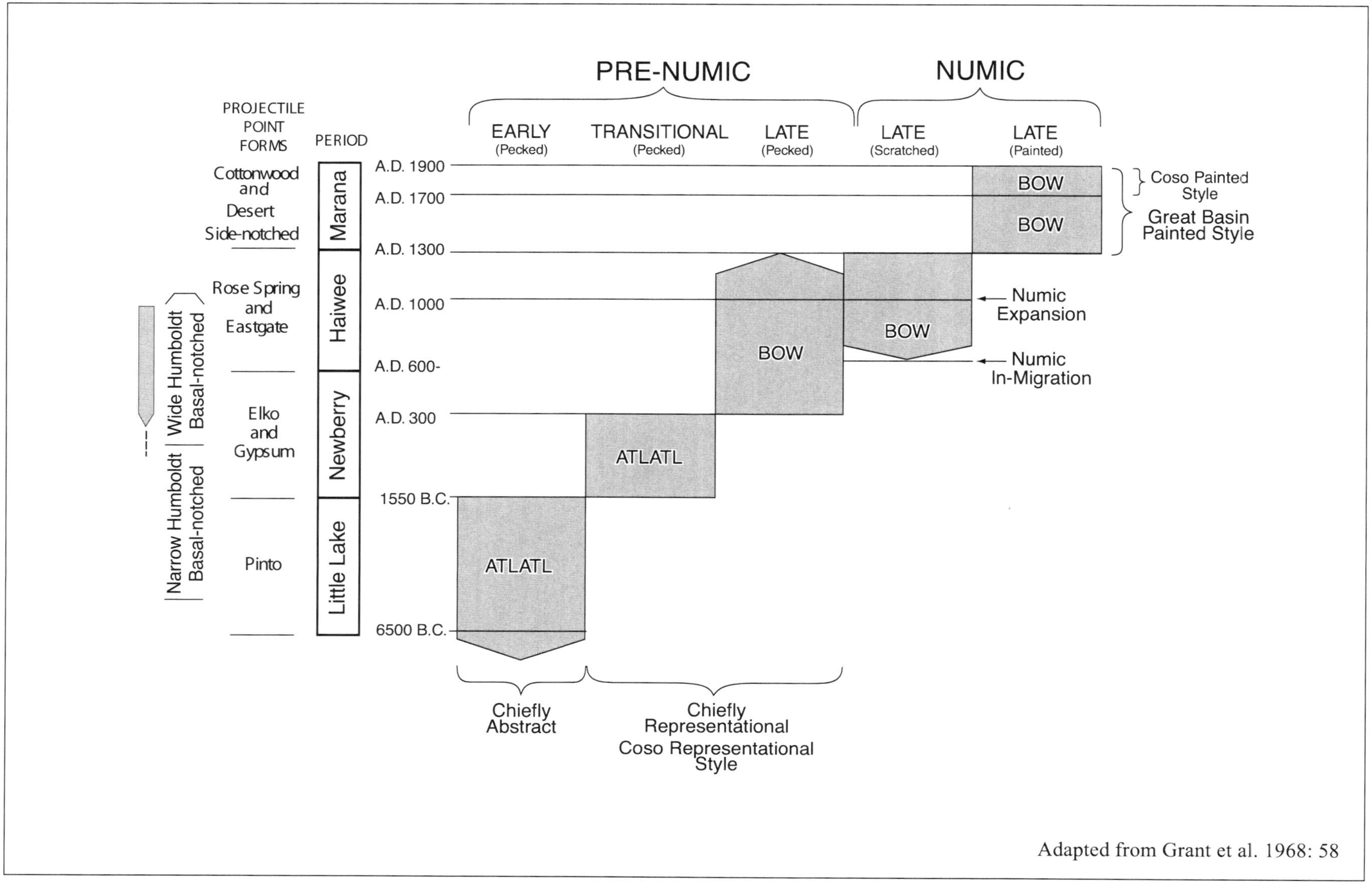

Figure 6.1 Dating of Weaponry, Rock Art Styles and Population Movements

(Bettinger and Baumhoff 1982; Heizer and Baumhoff 1962; Quinlan and Woody 2003; Nissen 1995). The impetus for the rise, fluorescence, and abandonment of this ornate rock art cannot be explained within the framework of a static model. A dramatic disjunction in the southwestern Great Basin archaeological record is marked by the demise of the Coso Representational Style and could represent the advent of an intrusive population. The latter would probably be the in-migration of Numic populations replacing pre-Numic groups. Such a massive stylistic turnover in material culture is, in fact, synonymous with the changes that could be predicted if a Numic expansion and population replacement were to have taken place (cf. Grayson 1994:23).

Catastrophic cultural conditions have been shown to correlate with revitalistic movements and an upsurge in ceremonialism (e.g., Coso petroglyph fluorescence). During the closing period of the Coso petroglyph tradition, such circumstances may have prevailed. An intruding and competitive population influx (Numic in-migration) and the dramatic depletion of the bighorn sheep could have factored into the demise of both the people and their artistic tradition (Garfinkel et al. 2005; Grant et al. 1967). These factors appear to be contemporaneous with the production spike and elaboration of Coso petroglyphs.

Numic and Pre-Numic Contemporaneity

Several Great Basin researchers posit a temporal overlap between Numic and pre-Numic occupations lasting a number of centuries (Bettinger 1994:53; Fowler and Madsen 1986; Garfinkel et al. 2004:94; Madsen 1986; Marwitt 1986; Young and Bettinger 1992; Zeanah 2002: 251). Logically, during the period of initial Numic migration and intrusion into eastern California, both groups (Numic and pre-Numic) must have occupied the region for some time simultaneously (cf. Hughes 1994; Madsen 1994).

Many eastern California prehistorians have also come to identify the beginning of the Haiwee Period, or ca. A.D. 600, as the most likely date for the initial population incursion by the Numic, and prior discussions (presented herein) support such a suggestion (Bettinger 1994; Delacorte 1994, 1995; Young and Bettinger 1992). Delacorte (1994) has, in fact, argued that there are differences in subsistence-settlement practices that allow prehistorians to differentiate Elko age pre-Numic settlements and nearly contemporaneous early Rose Spring Numic hunting camps.

If it were possible to discern the precise contemporaneity of particular settlements, such inferences could establish the synchronicity of two distinct cultural traditions (see discussions in Chapters 1 and 3). Unfortunately, such conditions are uncommon. Archaeologists have difficulty demonstrating the precise season and duration of settlement. The precision of chronological indicators also rarely allows confirmation that two sets of sites are exactly contemporaneous. There is considerable latitude with absolute dating techniques, such as radiocarbon assays, and even more so with obsidian dating. So many confounding factors impinging simultaneously may make the effort nearly hopeless. Yet in the present instance the hypothesis that such a pattern occurred within the study area appears worthwhile. A discussion is in order.

Two methods of piñon exploitation have been recognized (Dutcher 1893; Essene 1935). Brown-cone piñon nut procurement indicates the harvest of ripe nuts after the cones have opened in the fall (September through November). Under favorable conditions, the harvests using this collection technique were considerable (Bettinger and Baumhoff 1983; Simms 1985). However, competition from animals often restricted the length of the brown-cone harvest period. Given the well-documented, irregular productivity of piñon pines, these two factors combined to produce an exploitation pattern that was rather unreliable and ill-suited to the procurement and storage of large crops of nuts.

Alternatively, a green-cone piñon harvest method involved the exploitation of immature cones and the procurement of piñon pine cones before the cone scales had opened and the seeds dropped. This strategy involved collecting green cones from the trees by the use of a long, hooked stick and then roasting them to open the cone scales or, alternatively, caching the green cones in storage pits for future use. The green-cone method lengthened the period during which piñon pinenuts were harvested (August–November) and dramatically increased the yield of a seasonal piñon harvest.

Green-cone piñon harvesting resulted in different archaeological residues than brown-cone harvesting. The latter required little special processing or technology. However, intensive green-cone piñon pinenut procurement necessitated special technology and multiple stages of processing. Roasting pits, milling stones, and rock-ring piñon storage facilities were required for the harvest, preparation, and storage of unopened cones and loose nuts.

Most eastern California researchers recognize that green-cone piñon pinenut procurement is a rather recent technological introduction within the region. The majority of southwestern Great Basin prehistorians also believe that piñon was exploited only occasionally during the Newberry Period and earlier, using the nonintensive brown-cone method (Zeanah 2002). Such an interpretation is based on an absence of piñon cone roasting features or the charred remains of piñon scales and nut parts dating earlier than ca. A.D. 800 (Bettinger 1975; Eerkens and King 2002; Garfinkel and Cook 1979, 1981). Charred piñon nut remains have been recognized in archaeological deposits dating to the Newberry era, but they are uncommon. In the following Haiwee Period, a dramatic and noticeable increase in piñon archaeobotanical remains occurs. Rock ring, piñon cache features are amply documented beginning in the Haiwee Period. These archaeological manifestations clearly indicate the introduction of intensive green-cone piñon procurement (cf. Bettinger 1975; Bettinger and Baumhoff 1982; Hildebrandt and Ruby 1999).

Many Coso petroglyph sites are located in lowland settings at an average elevation of 5,000 feet amsl. They are often found near ideal hunting localities at entrances to gorges, on isolated rocks near springs, or along canyon walls above natural tanks. The latter collect and store water for long periods of time — even into the hot, dry, summer months (Gilreath 1999; Grant et al. 1968). Evidence that Coso petroglyph sites also functioned as ambush locations for bighorn hunts comes from the associated archaeological features including large numbers of hunting blinds and "dummy hunters" produced as stacked-rock features (Gilreath 1997, 2000:52; Grant et al. 1968; Muir 1901).

The survival of desert mountain sheep bands was critically dependent on access to predictable water sources, especially in the summer and early fall when sheep watered most frequently (Turner and Weaver 1981; Welles and Welles 1961). Bighorn traveled to natural tanks (*tinajas*) and springs during the late summer and early fall (August–September) for regular watering (Delacorte 1985, 1999; Madsen 1986). Studies of desert bighorn identify that limited home ranges (5 to 8 kilometers) are maintained and seasonal movements are regimented and predictable (Davis 1938; Geist 1971; Welles and Welles 1961). Geist (1971:79–81) notes that 75 to 90 per cent of the rams and ewes return to their home range following seasonal shifts. Bighorn behavior would therefore tend to make the sheep more susceptible to communal hunts (Pippin 1977).

In the summer, especially in dry regions where water sources are rare (such as the Coso region), bighorn hunters could hide near springs and natural tanks in order to trap these especially difficult-to-kill animals (Delacorte 1985; Steward 1933). Early historical accounts mention the practice of native peoples laying in wait as bands of sheep came to traditional watering holes (Bailey 1940; Nelson 1922; Muir 1901). Brook (1980) summarizes evidence for such bighorn hunting patterns and identifies half a dozen locations in the Coso Range where hunting blinds have been documented. All of these locations also feature Coso Representational petroglyphs.

Therefore, evidence for the inception of intensive green-cone piñon exploitation is recognized contemporaneously in the upland piñon forests of the Kern Plateau (6000 to 7500 feet amsl) and the Coso Range (above 7000 feet amsl). The occupation and use of lowland (5000 feet amsl on average) bighorn hunting sites are also associated with some of the largest collections of Coso petroglyphs (Figure 6.7). Piñon camps with numerous rock ring caches date to the beginning of the Haiwee/Sawtooth periods as do many of the Coso petroglyphs and hunting sites. That both green-cone piñon procurement and mountain sheep hunting were contemporaneous is also indicated by the substantial quantities of Rose Spring and Eastgate projectile points recovered at piñon camp sites in the upland piñon forests of the Cosos and the Kern Plateau Sierra crest (Hildebrandt and Ruby 1999; McGuire and Garfinkel 1980). The presence of obsidian hydration rims, Rose Spring and Eastgate projectile points, and rock art subject matter (realistically rendered arrow point images and hunters using bow and arrow) suggests that lowland Coso Petroglyph sites also date to this same time span (Garfinkel 2003; Garfinkel and Pringle 2004: Garfinkel et al. 2004; Gilreath 1999).

Pre-Numic peoples may have had difficulties simultaneously hunting mountain sheep and intensively harvesting green-cone piñon, an activity that requires a large number of individuals working together. Usually, multiple family groups, and often whole villages, would move together up to the piñon groves to share in the hard labor and elaborate efforts necessary (Steward 1938). There, in the groves, native people would set up temporary camps and harvest green cones. They would open the piñon cones on a sagebrush fire, free the nuts from their cone-scale homes, cache sufficient numbers to make the trip worthwhile, and pack back a nut crop large enough to cover subsistence needs for the participating families over the coming winter.

Madsen (1986) and Hildebrandt and Ruby (2000) argue strongly that scheduling difficulties might have occurred during the late summer and fall, (cf. Pippin 1977: Figure 4). A decision between which of the two strategies merited pursuit would have been a difficult one, owing to the costs and benefits of the two activities. It is difficult to reckon how such alternatives played out. Yet it is known that ultimately the balance shifted to green-cone piñon exploitation, in part perhaps because of overexploitation of ungulate populations (Garfinkel et al. 2005; Madsen 1986). That conclusion is supported by the late prehistoric (after A.D. 600) reduction in artiodactyl exploitation recognized in regional archaeofaunal assemblages (Basgall and Giambastiani 1995; Delacorte 1999).

The Coso region is small, yet it contains striking evidence of a highly standardized rock art tradition restricted to this area (Figures 6.2, 6.3, 6.4, 6.5 and 6.6). The bighorn hunting sites are less than 24 kilometers away from the upland Coso piñon forests. Both classes of sites are geographically coterminous with the territory exhibiting many of the unique hallmarks of Coso Representational Style petroglyphs (Figure 6.7).

As mentioned above, this core territory covers an area of less than 230 square kilometers. Yet, there are distinctive differences in the two sets of sites. Both sets appear to have been occupied at the same time. They could also represent different ethnic groups exploiting different aspects of the biotic environment (upland green-cone piñon and lowland communal bighorn hunts).

It is difficult to imagine how a small group of hunter-gatherers (50–100 people) could have been responsible for both sets of archaeological sites. This is all the more compelling given that both types of settlements are situated within such close range — within a few miles of one another (Figure 6.7). It is of course possible that different segments of the same cultural group were procuring upland green-cone piñon while others hunted lowland bighorn sheep. However, the labor requirements for piñon and the probable communal nature of Coso sheep hunts might lessen such a possibility.

It seems possible that the Numic were intensively using the upland piñon forests of the Cosos and the Kern Plateau during the Haiwee era. It is important to note that these same Numic groups were also using lowland localities for hunting small game and collecting hard seeds; see above. Only a small island of piñon exists in the Coso area. A sea of arid lands surrounds that small area. Hence subsistence decisions must have been highly restricted.

Recent studies in the Coso piñon area (Gilreath 2003) documented significant numbers of Great Basin Abstract rock art sites. Significantly, these early rock art sites date predominantly to either the Newberry or Little Lake periods, with only minor expressions of Haiwee-age Coso Style Representational petroglyphs (Gilreath 2003:213, Table 91). If a distinctly different cultural tradition (the Numic) were using the upland piñon forests for intensive, green-cone piñon harvests in contrast to groups occupying the lowland region for bighorn hunting during the Haiwee Period, a corollary manifestation of differing rock art iconography might be predicted.

Based on the exclusive superposition of scratched elements over pecked designs at the Coso Range upland piñon camps, Gilreath (2003) argues that these designs (Great Basin Scratched Style) are one of the more recent petroglyph styles in the western Great Basin. Other rock art researchers largely agree. They place the Great Basin Scratched Style recently in time beginning ca. A.D. 600 /1000 (Figures 6.1 and 6.2). That date would support the inception of Great Basin Scratched well after the termination of Great Basin Abstract but perhaps partially contemporaneous with the Great Basin Representational Petroglyph Styles, (e.g., Coso Representational Style). Further, the Great Basin Scratched glyphs appear to have ceased just prior to the inception of the Great Basin Painted Style (Heizer and Baumhoff 1962; Nissen 1982; Ritter 1994).

An extraordinary number of scratched drawings were documented at the Coso piñon forest sites. Nearly 300 individual scratched petroglyph elements were recorded at seven sites, with half of all documented rock art panels containing scratched elements (Figure 6.7). These scratched elements were in every instance either solitary designs or superimpositions over earlier Coso petroglyphs. Not a single scratched design was found beneath a Coso petroglyph design.

These rock drawings may be one of the largest concentrations of Great Basin Scratched Style rock art yet identified in the Great Basin. Many of these sites apparently date to the Haiwee Period based on their association with large numbers of Rose Spring points. Such a specific spatial distribution and chronological placement would seem to support the conclusion that the Numic used the Coso highlands for piñon nut procurement during the Haiwee Period. Another culturally distinct group (the pre-Numic) intermittently hunted bighorn sheep in the nearby lowland environs (cf. Ritter 1994).

Quinlan and Woody (2003) support just such a conclusion in asserting that Numic pioneer groups used scratched rock art to "socialize the landscape." These in-migrating Numic colonizers responded to pre-Numic rock art through modification of that art. They often placed their scratched designs in direct association with Coso petroglyphs, perhaps to obliterate, deface, or embellish them (cf. Quinlan and Woody 2003:372; Ritter 2004)

Cultural Succession and Rock Art Chronology

Figures 6.1 and 6.2 graphically display the rock art styles, associated weaponry, and cultural succession posited for the Coso Range and vicinity (also see Figures 6.3–6.6). On the right side of Figure 6.2 are hypothesized pre-Numic rock art expressions. It appears that, beginning around 8500 years ago, rock art production in the Coso Range began with imagery principally conforming to the Great Basin Abstract style. Those early panels included curvilinear meanders, circles, zigzags, chevrons, etc. Few realistic or naturalistic representations were rendered during this early period, but there were probably a few simple bighorn sheep (with horns to the side), atlatl designs, solid body anthropomorphs, and simple front-facing horn images.

Later, perhaps 3,000 or 4,000 years ago, a tradition began of more naturalistic and realistic designs, including more complex bighorn sheep, elaborate ceremonial figures known as "patterned-body anthropomorphs," and more embellished representations of atlatls. This tradition, which flourished, shows extensive continuity, development, and elaboration in subject matter and style from the earlier patterns.

The Coso drawings continued to develop with increasing complexity. Over time, petroglyphs were executed with more detail and in greater size, culminating with a pattern of larger-than-life-size sheep (some greater than 7 feet in length) done with great care. These drawings were rendered in unique Coso style with front-facing, bifurcated horns, boat-shaped bodies and flat backs with ears and hooves often added (see Figure 6.3 for a depiction of the changes in sheep renderings throughout the Coso tradition). The greatest number of these, associated with the Coso Representational Petroglyph tradition, apparently date to the Haiwee Period (ca. A.D. 600–1300). That expression appears to have ended abruptly, while in full fluorescence, during the Haiwee Period.

On the left side of Figure 6.2 are hypothesized Numic rock art styles. Numic rock art may have begun with very simple designs, known as the Great Basin Scratched Style, found throughout the Coso Range, in both the uplands (Coso Peak) and the lowlands (Little Lake, Upper and Lower Renegade Canyons, and Sheep Canyon), many times superimposed over pre-Numic glyphs (Figure 6.7). These rock art designs apparently initiated with the Numic in-migration, beginning as early as A.D. 600. They appear to have continued during the period when both Numic and pre-Numic occupations occurred in eastern California (ca. A.D. 600–1000).

The last and most recent manifestation of Numic rock art includes the infrequent occurrence of monochrome (red) abstract paintings known as an expression of the Great Basin Painted Style (Heizer and Clewlow 1973). A unique set of polychrome paintings labeled as Coso Style paintings (see Chapter 5) also occurs at the end of the Marana Period (A.D. 1300–Historic). The Coso Style paintings evidently date exclusively to the historic era and copy some of the imagery, style, and subject matter of the earlier Coso Representational petroglyphs. The element inventory of Coso Style paintings incorporates traditional subject matter, including bighorn sheep, atlatl- and spear- impaled animals and bow-and-arrow hunters (Garfinkel et al. 2005). Also appearing within the paintings are subjects of Euroamerican origin including hatted anthropomorphs, horses and rider, and cattle.

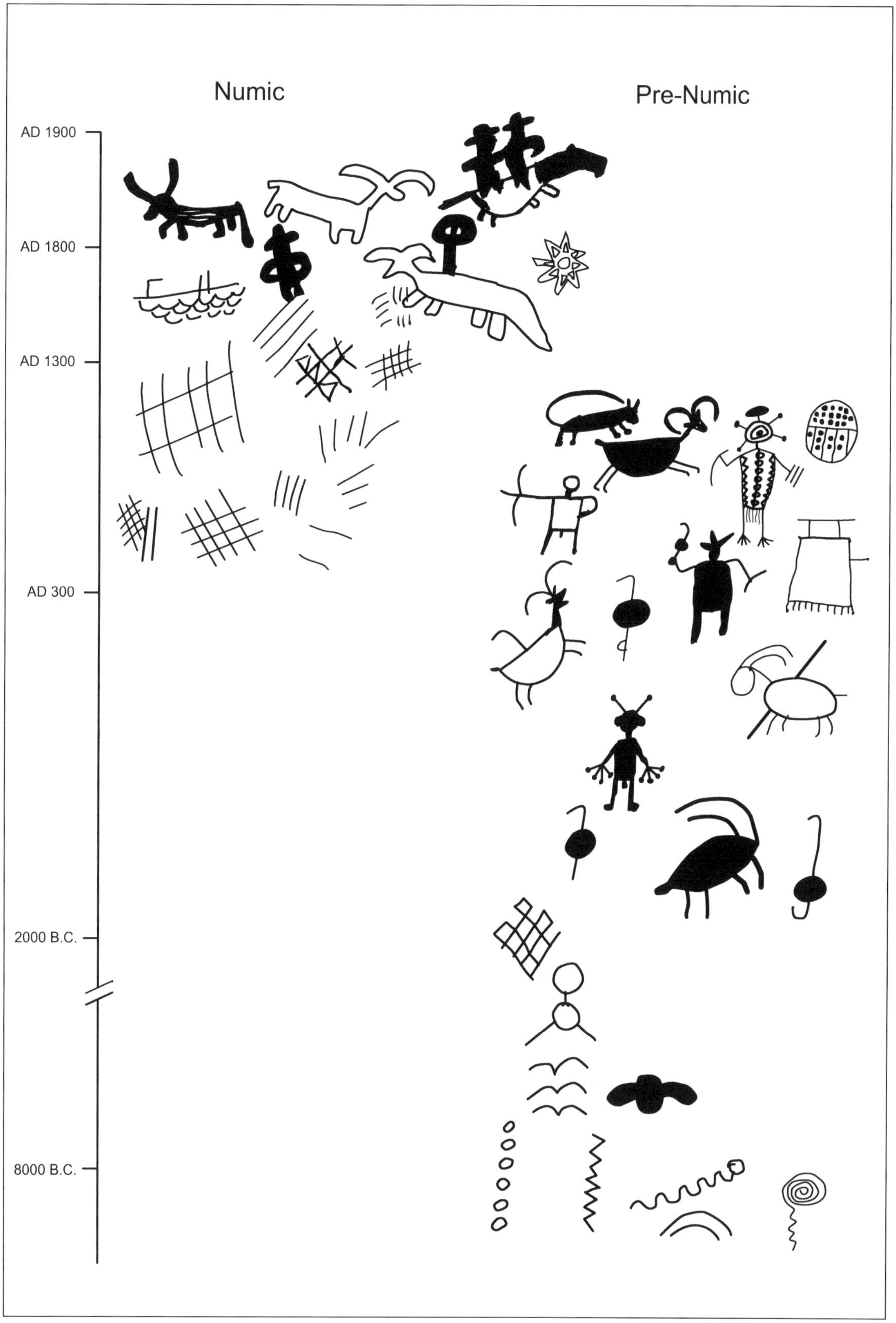

Figure 6.2 Kern Plateau and Coso Region General Rock Art Chronology.

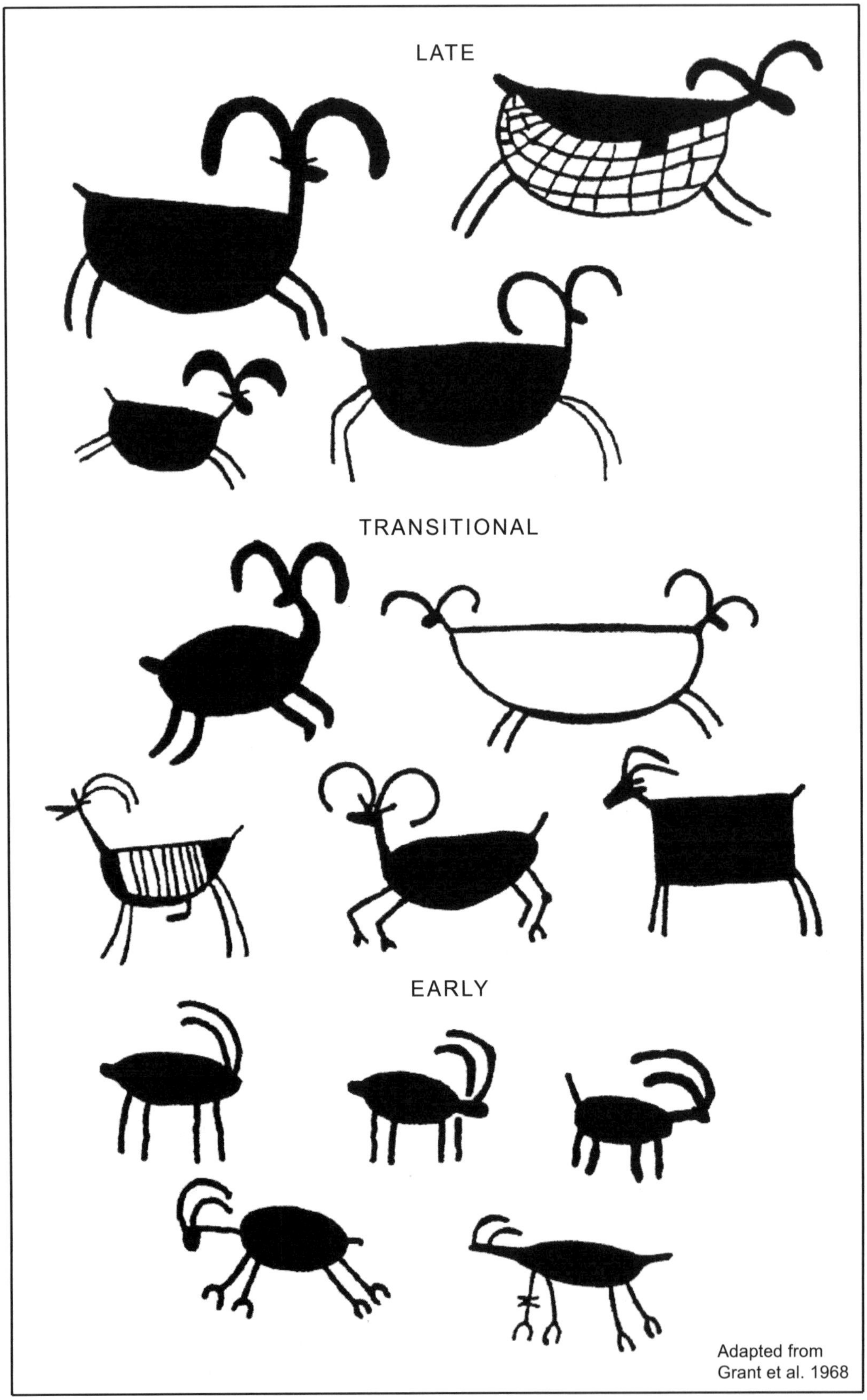

Figure 6.3 Characteristic Sheep Drawings and Chronological Periods.

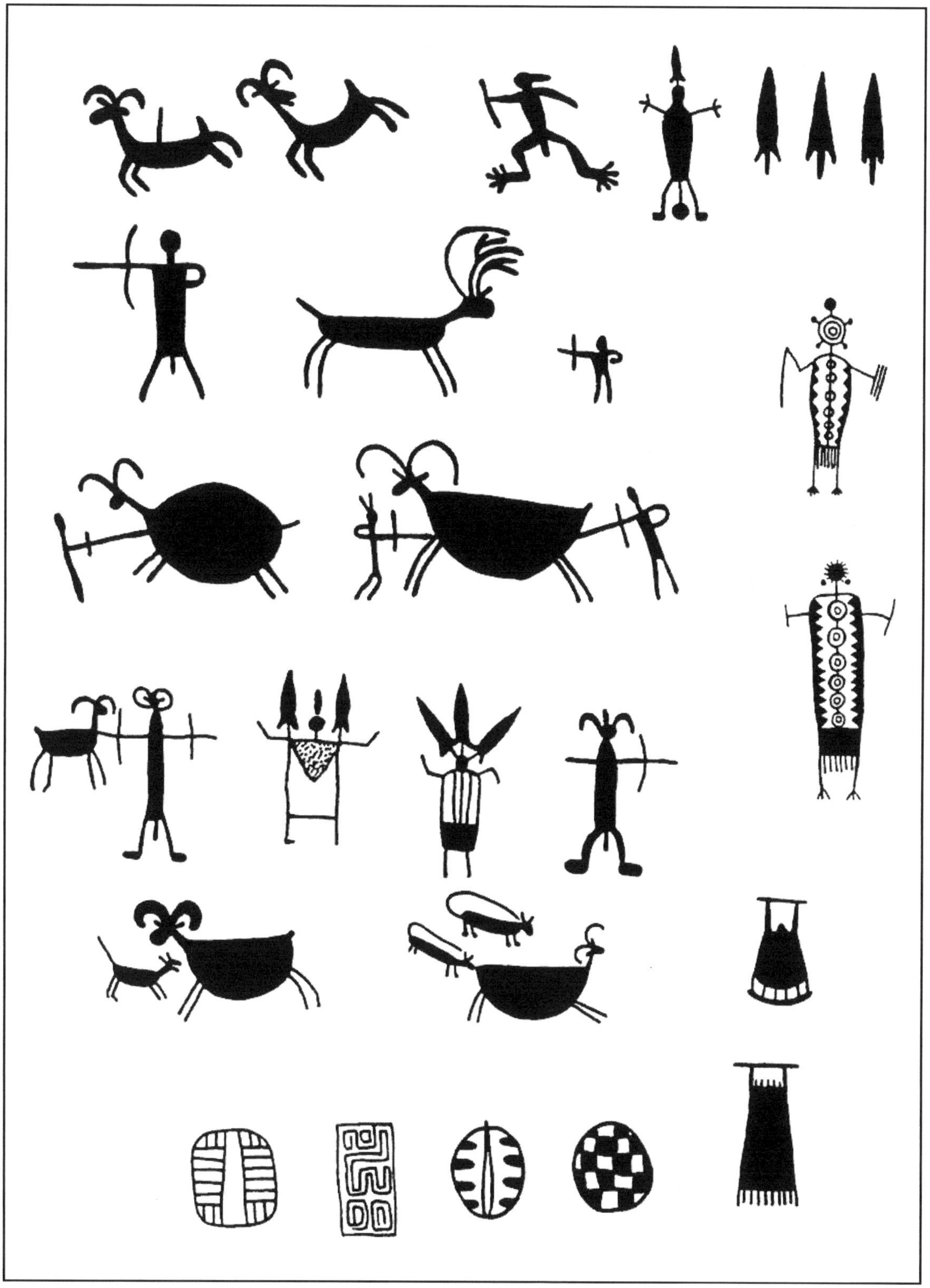

Figure 6.4 Late-Period Coso Petroglyph Elements.

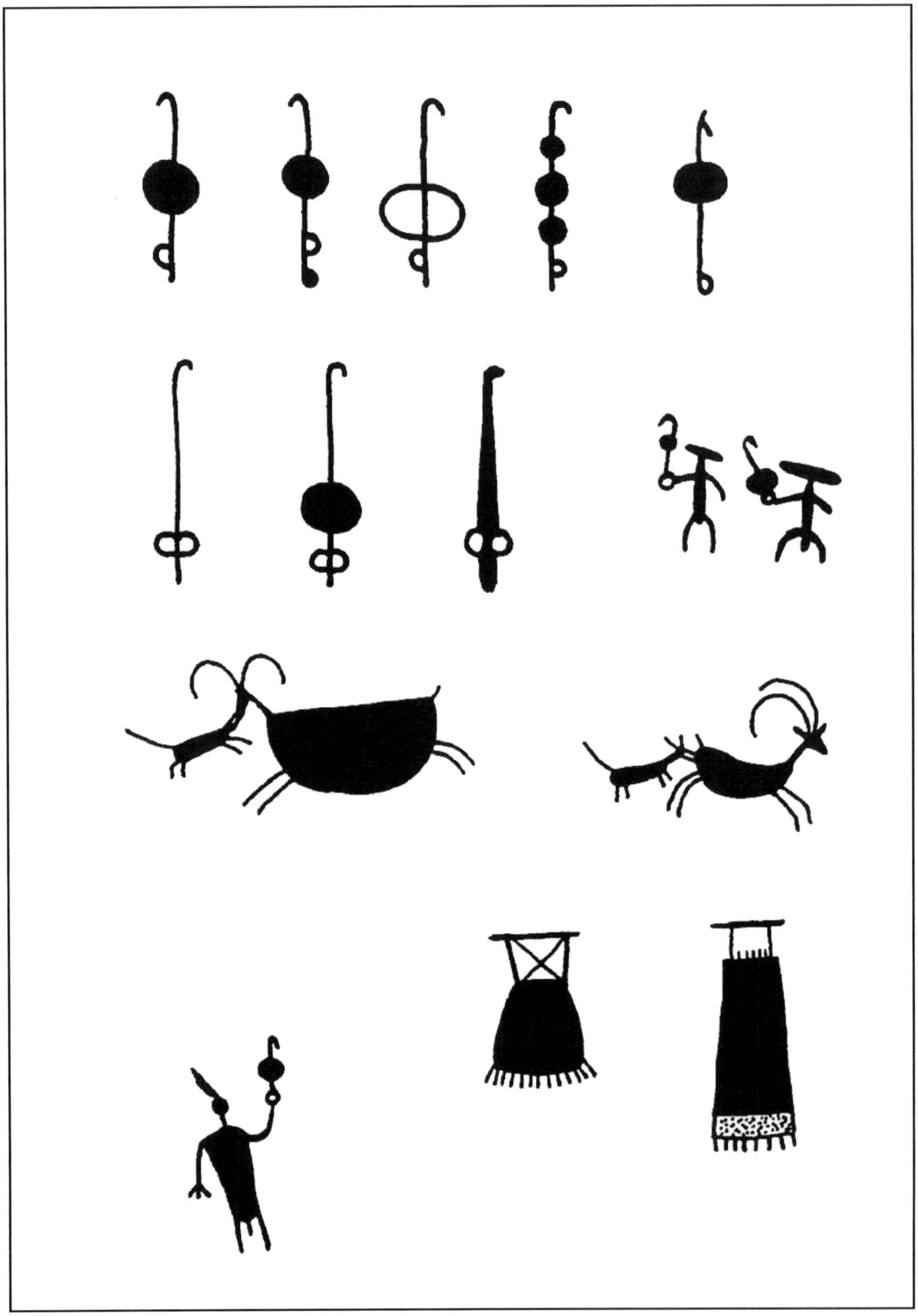

Figure 6.5 Transitional-Period Coso Petroglyph Elements.

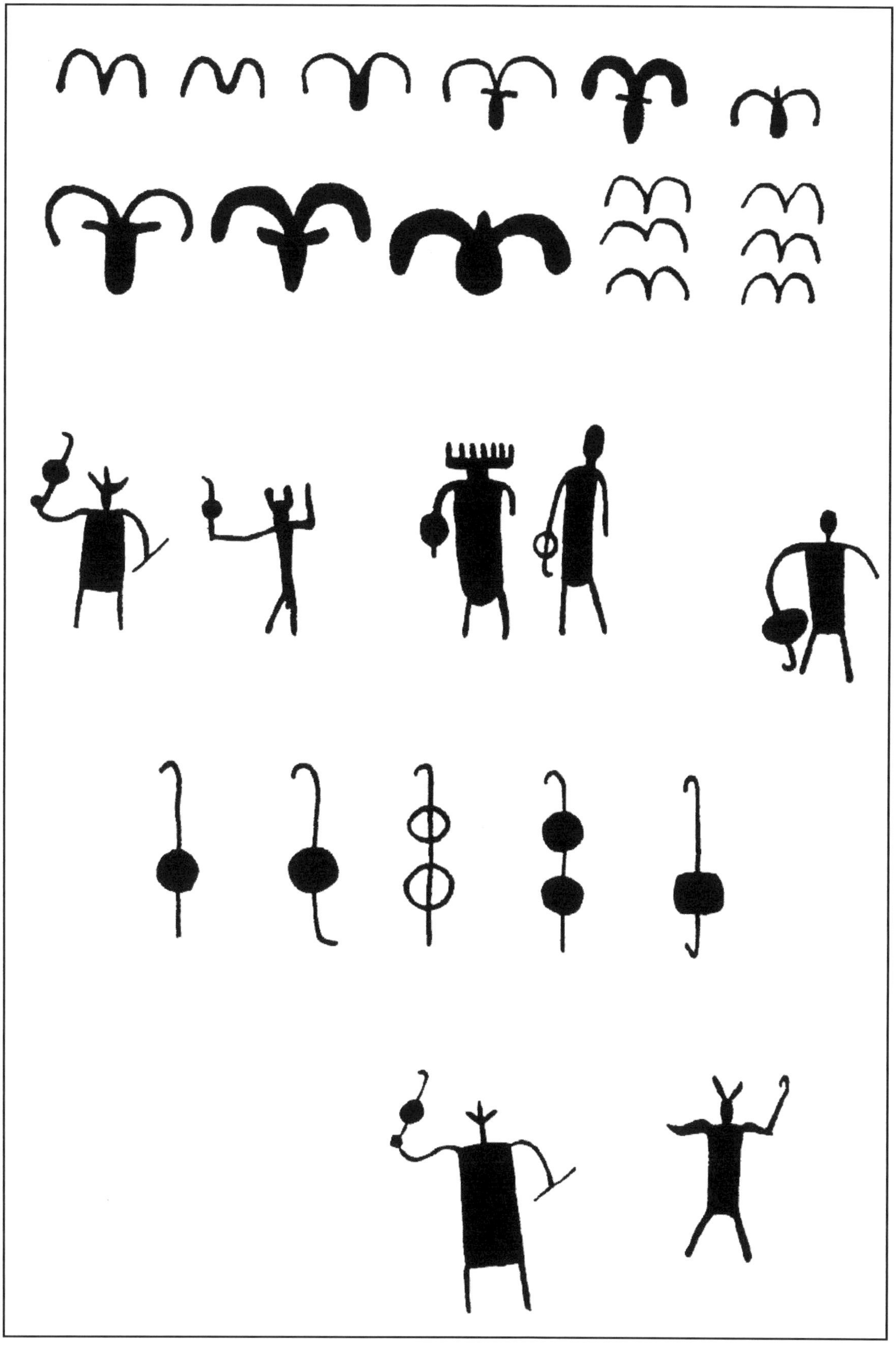

Figure 6.6 Early-Period Coso Petroglyph Elements.

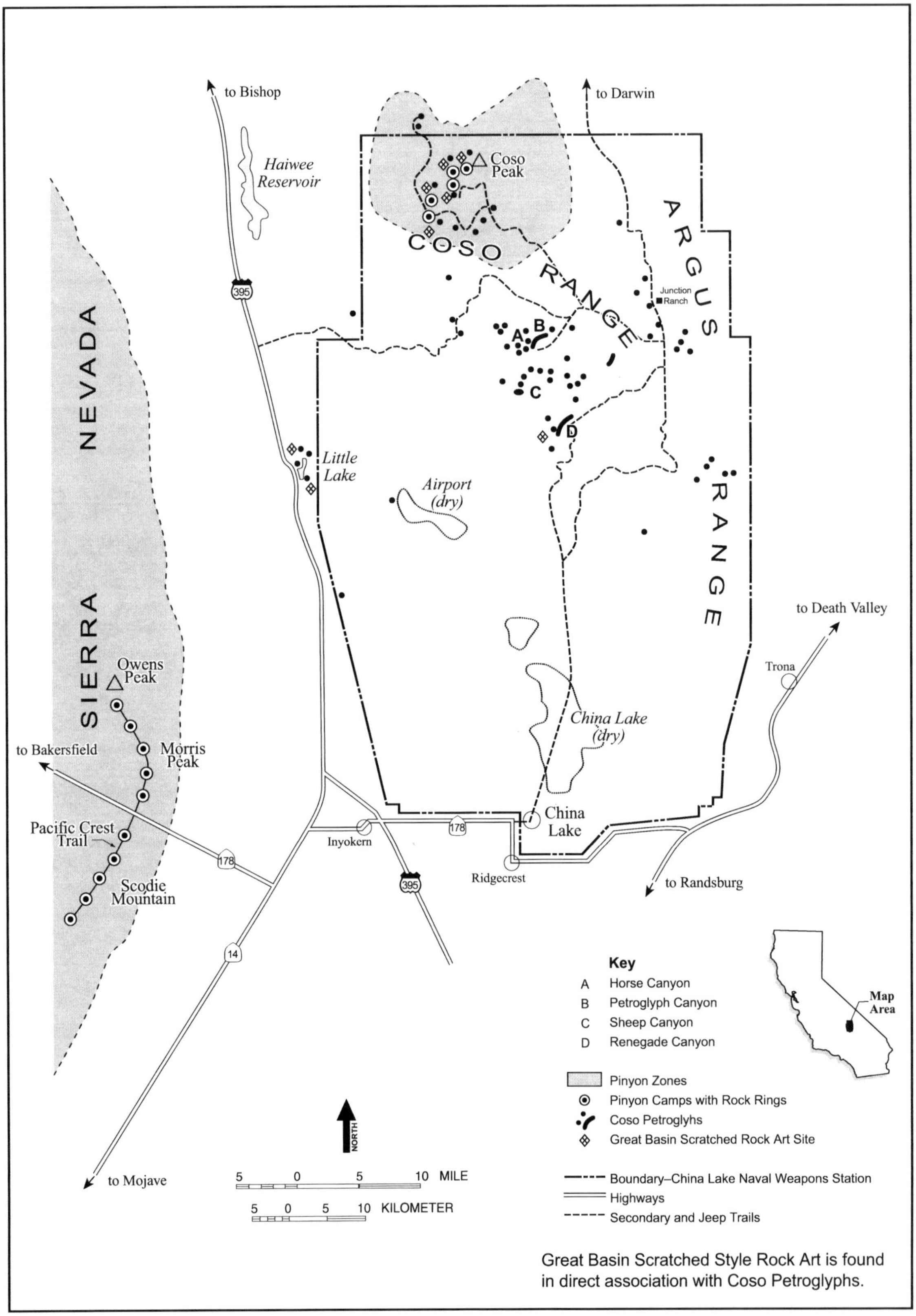

Figure 6.7 Coso Representational Petroglyphs, Piñon Zones, Piñon Camps, and Great Basin Scratched Rock Art.

Evaluation of Models of Numic Population Movements

In this final section, several models for Numic expansion (Chapter 3) are evaluated. Three main explanatory models have been proposed to account for the historic distribution of Numic groups within the Great Basin: Aiken and Witherspoon's *in-situ* cultural development, Sutton's conquest through conflict and warfare, and Bettinger-Baumhoff's economic displacement (cf. Delacorte 1995). The latter two models are not mutually exclusive but emphasize different aspects of the prehistoric adaptations in their suggestions of how the Numic spread may have taken place. These alternatives are considered here in light of the previous archaeological data and are presented with an eye toward critically reviewing their merits in the general study area.

Aikens-Witherspoon (in-Situ Development)

The Aikens and Witherspoon (1986:15; Aikens 1994) model proposes that Numic peoples have occupied the heart of the Great Basin since about 3500 years ago. Population-genetic studies in western Nevada seem to contradict that proposal, indicating a different pattern of mitochrondrial DNA than that characteristic of modern Numic groups. Rather than being a heartland for Numic ancestors, that area appears to have been an area of long-term pre-Numic occupation. Aikens also argues that the better-watered, wetland refugia on the peripheries of the Great Basin were actually the homes for pre-Numic rather than Numic groups. This prediction also is at odds with recent evidence for Numic populations in the northern Owens Valley based on population-genetic studies from Fish Slough Cave (Kemp et al. 2004). It is important to note that the model proposed by Aikens might be valid with reference to the ethnic groups on the fringes of the Great Basin, including the Washo, Fremont, and Anasazi.

Sutton (Conflict and Warfare)

Sutton (1986, 1987, 1991) proposes that Numic expansion was mired in conflict over the control of critical resource patches. He suggests that Numic invaders migrated into underused areas, taking control of crucial resource patches, and denying use to their non-Numic neighbors. Little evidence to support such scenarios is provided by the analyses reported in this book. Rather than denying use, pre-Numic groups appear to have only sparsely occupied certain areas with less intensive use of specific resources. Research has identified that, "the Numic were inhabiting very different ... sites than [were] their predecessors, a pattern...that helps explain how they could eventually displace existing groups." (Delacorte 1995:10)

It also seems that more intense procurement methods and labor-intensive adaptive poses are hallmarks of Numic subsistence-settlement strategies. Besides, pre-Numic peoples appear to have been characterized by low population densities, and that would seem to have precluded the type of intense competition or conflict proposed by Sutton (cf. Young and Bettinger 1991). Early in the Haiwee Period new settlements were directed toward areas of *lesser resource productivity,* not those areas previously favored. Many prehistorians would consider these new areas more marginal than the former, more central, settlements (cf. Delacorte 1994). In turn, based on optimal foraging models, the resources exploited by Numic people would ordinarily be ranked below first-order selections of certain high-caloric, lacustrine plants or large game animals. Resource exploitation by the Numic intruders was labor-intensive in terms of their heavy processing costs and lower return rates (e.g., grebes, lagomorphs, green-cone piñon nuts, and hard seeds).

Bettinger-Baumhoff (Economic Displacement)

Since the Bettinger-Baumhoff model of Numic expansion was largely derived from studies in eastern California, it is not surprising that the data presented here are largely consistent with its predictions. Limited prehistoric population-genetic studies have begun to support the position that Numic groups occupied the northern Owens Valley significantly earlier than in other parts of the Great Basin (e.g., western Nevada). Other genetic evidence also supports the contention that a population replacement occurred, at least in the western Nevada area of the Great Basin.

Yet if the Numic were always in the Owens Valley, it seems they went through several *in-situ* transformations. Bettinger (2002) has asserted that Owens Valley is clearly the source of the Numic spread. He further maintains that changes within Owens Valley were all *in-situ* transformational ones and that "*adaptive change everywhere else in the Great Basin is the result of migrations and population replacements from Owens Valley*" (Bettinger 2002). In other words, he believes that the Owens Valley is the proto-Numic homeland, and that the ancestral Numic were more ancient in this area than elsewhere in the Great Basin.

However, Bettinger seems to have concluded, at least during an earlier phase of his Owens Valley research, that pre-A.D. 600 expressions within the Owens Valley were most likely pre-Numic (Bettinger 1976:91–92, Table 4; Bettinger et al. 1984:125–129). He posits that a local population increase, through immigration of Numic groups into the Owens Valley, was partly responsible for the introduction of intensive green-cone piñon exploitation in the Inyo-White Mountains ca. A.D. 600, replacing the pre-Numic groups (Bettinger 1976). Bettinger et al. (1984) also identify a Newberry-age settlement (CA-Iny-2146, Partridge Ranch Site), located just south of Bishop in the central Owens Valley, as a "Pre-Numic site." They identified that site as having the archaeological materials characteristic of a "traveler strategy." In other words, a variety of elements expected at pre-Numic sites were represented there. These included a reliance on hunting and plant procurement emphasizing a low-cost strategy, a restricted range of plant exploitation and expansive land use evidenced by extensive tool curation and caching (Bettinger et al. 1984:126; cf. Bettinger and Baumhoff 1982).

The series of subsistence and settlement changes, perhaps marking adaptive shifts related to cultural succession, are in some cases precisely the same in the Owens Valley as outside this region (Delacorte 1990; Hildebrandt and Ruby 1999, 2003; McGuire and Garfinkel 1980). This pattern may suggest that a population replacement may also have occurred within the Owens Valley, as well as outside this region. This population in-migration, if it occurred, appears to have taken place at an earlier date. Specifically, shifts to small-game procurement, intensive plant exploitation, and changes in rock art style are all characteristic of the Owens Valley as well as some other areas of the Great Basin. These shifts may have taken place hundreds of years earlier in the Owens Valley than elsewhere in the Great Basin.

One class of data, widely accepted as evidence of Numic replacement is Great Basin Scratched rock art (Bettinger and Baumhoff 1982:494; Quinlan and Woody 2003). Yet these simple images might not be expected within the Owens Valley, if it had been occupied continuously by Numic peoples. However, both Bettinger and Baumhoff (1982:494) and von Werlhof (1965: Figures 26d, i, j, 27d, f, g, and others) identify at least 18 panels exhibiting just such scratched rock art superimposed over earlier Great Basin Abstract motifs located *within* Owens Valley. Such manifestations are puzzling and in need of further clarification.

As summarized previously, mitochondrial DNA evidence from Fish Slough Cave supports Numic occupation in the Owens Valley from at least A.D. 800. However, Nelson (1999:189) emphasizes that her studies at Fish Slough Cave provide evidence for occupation by "different populations." That evidence includes the presence of both Numic and non-Numic types of basketry and the occurrence of raw material in cordage that was not of Numic affiliation. These data are far from compelling but provide tantalizing indications of pre-A.D. 600 occupation in the Owens Valley by pre-Numic groups.

Summary

In the Kern Plateau interior, archeological evidence confirms the conclusion that the Tubatulabal language and cultural tradition are of long standing. Several sites show continuous, unbroken occupation from the historic era back 2500 years or more. Distributions of obsidian hydration rim measurements indicate a relatively uninterrupted prehistoric sequence. Dietary patterns also show a consistent emphasis on large game hunting and pinenut use over two millennia. The use of a single geographic source of volcanic glass and the rendition of a solitary rock art tradition further testify to a single cultural expression.

In the desert region to the east, burial patterns at Chapman Cave suggest a direct historical link between ethnographic Numic populations and their earlier archaeological manifestations from ca. A.D. 600. Mitochondrial DNA evidence from the northern Owens Valley supports Numic occupation from ca. A.D. 800, skeletal materials from western Nevada imply a recent population expansion by Numic groups, and a replacement of pre-Numic groups within the last 500 years (ca. A.D. 1300).

The Sierra Nevada crest and the eastern California desert areas witnessed significant subsistence shifts beginning ca. A.D. 600 and an evident disjunction in the midst of the Haiwee Period. These changes include a decline in hunting of large game, an initial and growing emphasis on dryland hard seeds, the beginning of intensive green-cone piñon pinenut use, and the mass harvest of easily procured and abundant small game animals. Obsidian hydration distributions indicate the timing and intensity of these shifting occupation patterns. These changes may reflect Numic in-migration and expression of their distinctively different adaptations.

Earlier pre-Numic settlements appear to have emphasized obsidian acquisition, ungulate hunting, and rock art production. An abrupt cessation of the Coso Representational rock art tradition and its replacement by a simple scratched style is believed to have signaled disruption of autochthonous groups by an exotic population. Alternative models of Numic population movements were compared and the Bettinger-Baumhoff model of economic displacement appears to best fit the available data. Nonetheless, several puzzling data sets still support the idea that an early, pre-A.D. 600, pre-Numic occupation may have existed in the Owens Valley.

Chapter 7
Conclusions

This brief chapter sets forth this book's major conclusions. Archeological data support the hypothesis that the Tubatulabal cultural tradition is of long standing in the South Fork Valley of the Kern River area (Isabella Basin) and the interior Kern Plateau. A variety of archaeological evidence supports a continuous, unbroken occupation from the historic era back 2500 years or more. Distributions of obsidian hydration rim values, subsistence data, trends in obsidian source use and acquisition, and representation of a single rock art tradition uniformly indicate that a single cultural expression lasted several millennia.

Archaeological evidence also suggests that initial pre-Numic colonizers settled in the well-watered lowland areas of the adjacent desert areas to the east leaving large expanses of land relatively open. This pre-Numic subsistence-settlement pattern apparently focused on highly ranked (high caloric) foods with low processing costs (such as large artiodactyls, brown-cone piñon, and riparian resources). During the Little Lake and Newberry periods, such residential occupation sites can be found at Little Lake (the Stahl Site and Stahl Site Cave), Rose Spring, Portuguese Bench, and Lubkin Creek.

During the subsequent Haiwee Period, dual use of the landscape may have taken place. Pre-Numic residential occupations appear to have continued, but Numic people also may have begun to target a different series of resources at around A.D. 600. Numic sites were positioned so that people could exploit resource-rich patches, while minimizing travel time. Such sites can be recognized in the upland piñon camps of the Sierra Crest and Coso Range, the seed camps in the lowland Coso Volcanic Fields, a grebe-hunting camp at Ash Creek near Owens Lake, and a jackrabbit-hunting settlement in the El Paso Mountains, among many other similar sites. Initial Numic in-migration may be attested to by rock art of the Great Basin Scratched Style.

Haiwee Period Numic hunting camps could owe their brief existence to a number of unusual factors that took place only during the Haiwee Period (A.D. 600–1300). One factor would be that non-Numic and Numic groups possibly coexisted for several hundred years and exhibited different subsistence patterns. Additionally, at least two periods of intense drought (1058–838 B.P. and 741–600 B.P.), termed the Medieval Climatic Anomaly (MCA) (Stine 1994; see also Jones et al. 2004), appear to have influenced resource decisions during the late Haiwee Period.

Occupations at pre-Numic residential sites appear to have continued into the early Haiwee Period. Apparently such sites declined in use, and in some areas seem to have been abandoned altogether, toward the end of that period. This discontinuity is evidenced by an abrupt decline in Coso lowland hydration readings from the Rose Spring, Portuguese Bench, and Coso Volcanic Field sites. At about this same time, other cultural changes took place. The manufacture of Coso Representational petroglyphs abruptly ceased at the height of their elaboration. Similarly, the production of Wide Humboldt Basal-notched bifaces was discontinued at the peak of their popularity.

During the following Marana Period, it appears that the pre-Numic people either became extinct (cf. Young and Bettinger 1992) or were absorbed into the more successful Numic populations. Limited genetic evidence suggests that a population replacement may have occurred in western Nevada (Kaestle and Smith 2001). That pattern may also be characteristic of the study area, but if it did occur it would have taken place considerably earlier. Genetic evidence from the northern Owens Valley is consistent with archaeological studies suggesting continuous Numic occupation from about A.D. 600.

In the Marana Period, the Numic people may have occupied lowland village sites that the pre-Numic people colonized earlier. Numic groups, at the same time, seem to have abandoned their short-lived and specialized Haiwee Period hunting camps. The Stahl Site Cave and the Rose Spring Site exhibit an early occupation during the Newberry Period, a hiatus or greatly diminished cultural activities during much of the Haiwee Period, and then an apparent reoccupation and intensification during the Marana Period.

The weight of evidence suggests that a Numic population incursion was in part responsible for the archaeological record in portions of eastern California and the far southern Sierra Nevada Crest. Some researchers see continuity between the historic Numic occupants and some of the more ancient archaeological manifestations in the region. This is especially the case with respect to the realistic petroglyphs recorded on the lava cliffs and canyons of the Coso Range. Yet, the body of evidence, when reviewed in detail and considered contextually, strongly indicates otherwise.

Contrary to the naysayers, a static archaeological record, or one with an especially strong sense of continuity is not to be found in much of eastern California. Settlements are punctuated with abrupt changes. Brief, and even lengthy, periods of site abandonment are common, and there are clear disjunctions in the archaeological record.

An *in situ* population increase would hardly have been predicted for this area given its environmental correlates. The Coso Range, Rose Valley, Indian Wells Valley, and far southern Sierra were used only selectively prior to the Haiwee Period. These areas are resource-deficient and arid and would present little in the way of natural resources to support elevated population densities. Hence, *in situ* population growth hardly seems sufficient to account for the abrupt shift in subsistence-settlement and technological changes within the span of a few hundred years.

A Numic population influx into eastern California would have resulted in territorial shifts exacerbating increased population pressure and reducing territory size. The Numic immigration appears to have immediately preceded the MCA. MCA droughts could have reduced biotic productivity and required new subsistence activities. The periods of aridity also could have spurred subsistence intensification in eastern California and led to the Numic replacement of pre-Numic groups and the ultimate expansion of Numic people out into the rest of the Great Basin.

Archaeological evidence has been advanced to evaluate alternative scenarios for aboriginal population movements in the southwest Great Basin. Population genetic studies are silent on the issue with respect to the premise that Numic groups were extant during the pre-Haiwee era in eastern California. This conclusion is not necessarily false; but there is no compelling reason to believe it is true. Given current evidence, the most plausible scenario is that a Numic migration and replacement of older population(s) occurred in the desert portion of the study area.

Prehistoric research in the Kern Plateau and Isabella Basin supports the view that the Tubatulabal cultural tradition is long standing–with relatively continuous occupation dating from historic times back 2500 years or more. Detailed examination of the archaeological record, along the crest of the far southern Sierra and in eastern California, supports the position that an influx of Numic groups took place around A.D. 600. Synchronous with the introduction of the bow and arrow, the Numic pioneers appear to have established themselves in the southwestern Great Basin. Numic populations apparently pursued a more successful adaptive strategy, either replacing or absorbing pre-Numic groups. Numic peoples subsequently expanded out of this heartland and migrated throughout much of the Great Basin.

References

Adovasio, J. M.

1986 Artifacts and Ethnicity: Basketry as an Indicator of Territoriality and Population Movements in the Prehistoric Great Basin. In *Anthropology of the Desert West: Essays in Honor of Jesse D. Jennings*, C. J. Condie and D. D. Fowler, editors, pp. 43-89. University of Utah Anthropological Papers No. 110. Salt Lake City: University of Utah Press.

Aikens, C. M.

1994 Adaptive Strategies and Environmental Change in the Great Basin and Its Peripheries as Determinants in the Migrations of Numic-Speaking Peoples. In *Across the West: Human Population Movement and the Expansion of the Numa*, D. B. Madsen and D. Rhode, editors, pp. 35-43. Salt Lake City: University of Utah Press.

Aikens, C. M. and Y. T. Witherspoon

1986 Great Basin Numic Prehistory: Linguistics, Archaeology, and Environment. In *Anthropology of the Desert West: Essays in Honor of Jesse D. Jennings*, C. Condie and D. D. Fowler, editors, pp. 7-20. University of Utah Anthropological Papers No. 110. Salt Lake City: University of Utah Press.

Allen, M. W.

1986 *The Effects of Bow and Arrow Technology on Lithic Production and Exchange Systems: A Test Case Using Debitage Analysis.* M.A. Thesis, Department of Anthropology, University of California, Los Angeles.

Ambro, R. D., M. K. Crist, B. J. Peck, D. M. Varner, and A.V. Eggers

1981 *Archaeological Investigations for the Scodie Mountains Segment of the Pacific Crest Trail, Kern County, California.* Report on file at the United States Department of Agriculture, United States Forest Service, Sequoia National Forest, Porterville, California.

Amsden, C. M.

1937 The Lake Mojave Artifacts. In The Archaeology of Pleistocene Lake Mojave, E. W. C. Campbell, W. H. Campbell, E. Antevs, C. A. Amsden, J. A. Barbieri and F. D. Bode, editors, pp. 51-97. *Southwest Museum Papers* No. 11. Los Angeles.

Andrews, S. B.

1977 Pictographs of the Tubatulabal. *Kern County Archaeological Society Journal* 1:33-42.

1980 Pictographs. In *Archaeological Investigations in the Southern Sierra Nevada: The Lamont Meadow and Morris Peak Segments of the Pacific Crest Trail,* A. P. Garfinkel, R. A. Schiffman and K. R. McGuire, editors, pp. 326-347. Cultural Resources Publications, Archaeology. Department of the Interior, Bureau of Land Management, Bakersfield District, Bakersfield.

Antevs, E.

1952 Climatic History and Antiquity of Man in California. *University of California Archaeological Survey Reports* 16:23-31.

Arkush, B. S.

1995 The Archaeology of CA-MNO-2122: A Study of Pre-Contact and Post-Contact Lifeways Among the Mono Basin Paiute. *University of California Anthropological Records* 31. Berkeley.

Austin, D.

2005 *The Stahl Site Petroglyphs: New Observations and Comments.* Article electronically published and accessed on June 25, 2005, at *www.petroglyphs.us/article_stahl_site_petroglyphs.htm.*

Backes, Jr., C. J.

2004 More Than Meets the Eye: Fluorescence Photography for Enhanced Analysis of Pictographs. *Journal of California and Great Basin Anthropology* 24(2):193-206.

Bailey, V.

1940 Into Death Valley 50 Years Ago. *Westways* 32(2):8-11.

Barbour, M. G. and W. D. Billings

2000 *North American Terrestrial Vegetation.* Cambridge: Cambridge University Press.

Barbour, M. G., T. Keeler-Wolf, and A Schoenherr

2006 *Terrestrial Vegetation of California*. W. C. Press.

Barbour, M. G. and J. Major

1977 *Terrestrial Vegetation of California*. New York: Wiley.

1988 *Terrestrial Vegetation of California*. Sacramento: California Native Plant Society.

Bard, J., C. I. Busby, and R. J. Dezzani

1985 *Kennedy Meadows Cultural Resources Data Recovery*. Report on file at the United States Department of Agriculture, United States Forest Service, Sequoia National Forest, Porterville, California.

Barras, J.

1973 *Tehachapi, the Formative Years*. Private Printing.

1976 *The Long Road to Tehachapi*. Bakersfield: Sierra Printers.

1984 *Their Places Shall Know Them No More*. Tehachapi: Privately Printed by Judy and Bud Barras.

Barth, F. (editor)

1969 *Ethnic Groups and Boundaries*. Boston: Little, Brown and Company.

Basgall, M. E.

1982 Archaeology and Linguistics: Pomoan Prehistory as Viewed from Northern Sonoma County, California. *Journal of California and Great Basin Anthropology* 4(1):3-22.

1988 Archaeology of the Komodo site, an Early Holocene Occupation in Central-eastern California. In Early Human Occupation in Western North America: the Clovis-Archaic Interface, J. Willig, C. M. Aikens and J. Fagan, editors, pp. 103-119. *Nevada State Museum Anthropological Papers* 21.

1989 Obsidian Acquisition and Use in Prehistoric Central-Eastern California: A Preliminary Assessment. In Current Directions in California Obsidian Studies, R. E. Hughes, editor, pp. 111-126. *Contributions of the University of California Archaeological Research Facility* 48. Berkeley.

1990 Hydration Dating of Coso Obsidian: Problems and Prospects. Paper presented at the 24th Annual Meeting of the Society for California Archaeology, Foster City.

1993 *Early Holocene Prehistory of the North-Central Mojave Desert*. Ph.D. Dissertation, Department of Anthropology, University of California, Davis.

Basgall, M. E. and M. G. Delacorte

2003 *Phase II Evaluations at Nine Archaeological Sites Near Independence, Inyo County, California*. Report on file California Department of Transportation, Fresno. ARC Technical Report 02-59.

Basgall, M.E., and M. Giambastiani

1995 Prehistoric Use of a Marginal Environment: Continuity and Change in Occupation of the Volcanic Tablelands, Mono and Inyo Counties, California. *Center for Archaeological Research at Davis Publication* 12.

Basgall, M. E and M. C. Hall

2000 Morphological and Temporal Variation in Bifurcate-Stemmed Dart Points of the Western Great Basin. *Journal of California and Great Basin Anthropology* 22(2):237-276.

Basgall, M. E. and W. R. Hildebrandt

1980 Faunal Remains. In *Archaeological Investigations in the Southern Sierra Nevada: The Lamont Meadow and Morris Peak Segments of the Pacific Crest Trail*, A. P. Garfinkel, R. A. Schiffman and K. R. McGuire, editors, pp. 288-316. Cultural Resources Publications, Archaeology. Department of the Interior, Bureau of Land Management, Bakersfield District, Bakersfield.

Basgall, M. E. and K. R. McGuire

1988 *The Archaeology of CA-Iny-30: Prehistoric Culture Change in the Southern Owens Valley, California*. Report on file, California Department of Transportation, District 09, Bishop.

Bass, S. T. and S. B. Andrews

1977 Using a New Classification System on Kern County Indians Beads. *Kern County Archaeological Society Journal* 1:9-24.

Baumhoff, M. A.

1957 An Introduction to Yana Archaeology. *University of California Archaeological Survey Reports* 40:1-61. Berkeley.

Baumhoff, M. A. and J. S. Byrne

1959 Desert Side-notched Points as a Time-marker in California. *University of California Archaeological Survey Reports* 48(72):32-65. Berkeley.

Bellwood, P.

1985 *Prehistory of the Indo-Malaysian Archipelago.* Sydney: Academic Press

1997 Prehistoric Cultural Explanations for Widespread Linguistic Families. In *Archaeology and Linguistics: Aboriginal Australia in Global Perspective,* McConvell and N. Evans, editors, pp. 13-134. Melbourne: Oxford University Press.

Bennyhoff, J. A.

1994 Central California Augustine: Implications for Northern California Archaeology. In Toward a New Taxonomic Framework for Central California Archaeology, edited by R. Hughes. *Contributions to the University of California Archaeological Research Facility* 52. Berkeley.

Bennyhoff, J. A. and R. E. Hughes

1987 Shell Bead and Ornament Exchange Networks between California and the Western Great Basin. *Anthropological Papers of the American Museum of Natural History* 64(2). New York.

Bettinger, R. L.

1975 *The Surface Archaeology of Owens Valley, Eastern California: Prehistoric Man-land Relationships in the Great Basin.* Ph.D. Dissertation, Department of Anthropology, University of California, Riverside.

1976 The Development of Pinyon Exploitation in Central Eastern California. *Journal of California Anthropology* 3(1):81-95.

1977 Aboriginal Human Ecology in Owens Valley: Prehistoric Change in the Great Basin. *American Antiquity* 42(1):3-17.

1979 Multivariate Statistical Analysis of a Regional Subsistence-Settlement Model for Owens Valley. *American Antiquity* 44(3):455-470.

1982 Aboriginal Exchange and Territoriality in Owens Valley, California. In *Contexts for Prehistoric Exchange,* J. E. Ericson and T. K. Earle, editors, pp. 103-127. New York: Academic Press.

1989 The Archaeology of Pinyon House, Two Eagles, and Craters Middens: Three Residential Sites in Owens Valley, Eastern California. *Anthropological Papers of the American Museum of Natural History* Number 67, New York.

1991 *Hunter-Gatherers: Archaeological and Evolutionary Theory.* New York: Plenum.

1994 How, When, and Why Numic Spread. In *Across the West: Human Population Movement and the Expansion of the Numa,* D. B. Madsen and D. Rhode, editors, pp. 44-55. Salt Lake City: University of Utah Press.

1999 What Happened in the Medithermal? In *Models for the Millennium: Great Basin Anthropology Today,* C. Beck, editor, pp. 62-74. Salt Lake City: University of Utah Press.

2002 *How Hunter Gatherers Spread.* Article electronically published and accessed on October 25, 2005, at *http://crlc.anu.edu.au/arcling2/Bettinger.htm.*

Bettinger, R. L. and M. A. Baumhoff

1982 The Numic Spread: Great Basin Cultures in Competition. *American Antiquity* 47(3):485-503.

1983 Return Rates and Intensity of Resource Use in Numic and Pre-Numic Adaptive Strategies. *American Antiquity* 48(4):830-834.

Bettinger, R. L., M. Delacorte, and K. R. McGuire

1984 *Archaeological Excavations at the Partridge Ranch Site (CA-INY-2146), Inyo County, California.* Report on file, California Department of Transportation, District 9, Bishop, California.

Bettinger, R. L., and R. E. Taylor

1974 Suggested Revisions in Archaeological Sequences of the Great Basin in Interior Southern California. *Nevada Archaeological Survey Research Paper* No. 5:1-26.

Binford, L. R.

1968 Archaeological Perspectives. In *New Perspectives in Archaeology*, S. R. Binford and L. R. Binford, editors, pp. 5-32. Chicago: Aldine.

1986 An Alywara Day: Making Men's Knives and Beyond. *American Antiquity* 51(3):547-562.

Blalock, J. M.

1979 *Social Statistics*. 2nd Edition. New York: McGraw Hill Company.

Boyd, W. H.

1972 *A California Middle Border, The Kern River Country, 1772-1880*. Richardson, Texas: The Havilah Press.

Brinton, D. G.

1891 *The American Race: A Linguistic Classification and Ethnographic Description of the Native Tribes of North and South America*. New York: N. D. C. Hodge.

Brook, R. A.

1980 Inferences Regarding Aboriginal Hunting Behavior in the Saline Valley, Inyo County, California. *Journal of California and Great Basin Anthropology* 2(1):60-79.

Brook, R. A., E. Ritter and N. Farrell

1977 Native American Rock Art in a 19th Century California Mining Boom Town. In *American Indian Rock Art* Volume IV, Papers Presented at the Fourth Annual A.R.A.R.A. Symposium, E. Snyder, A. J. Bock and F. Bock, editors, pp. 9-20. El Toro: American Rock Art Research Association.

Buechner, H. K.

1960 The Bighorn Sheep in the United States, Its Past, Present, and Future. *Wildlife Monographs* 4.

Burke, T. D., J. A. Markos, and L. A. Walsh

1995 *Tibbi Opo: The Prehistory of CA-Iny-4646, Northern Owens Valley, Inyo County, California*. Report on file Eastern Information Center, University of California, Riverside.

Burton, J. F.

1996 Three Farewells to Manzanar: The Archaeology of Manzanar National Historic Site, California. Western Archaeological and Conservation Center, National Park Services, *United States Department of Interior Publications in Anthropology* 57.

Butterbredt, I.

1948 *Kern County Indians*. Unpublished manuscript prepared during investigation by the California Archaeological Survey. Mimeographed. Manuscript on file, Southern San Joaquin Archaeological Information Center, California State University, Bakersfield.

Byrd, B. F. and S. N. Reddy

2004 *Phase II Archaeological Testing and Evaluation of CA-Iny-3647, Ca-Iny-3650/H, CA-Iny-3826, and P-14-7356, North Little Lake Rehabilitation Project, U. S. 395, Inyo County California*. Report on file, California Department of Transportation, District 9, Bishop, California.

Campbell, E. W. C. and C. A. Amsden

1934 The Eagle Mountain Site. *Masterkey* 8(6):170-173.

Campbell, E. W. C. and W. H. Campbell

1935 The Pinto Basin Site: An Ancient Aboriginal Camping Ground in the California Desert. *Southwest Museum Papers* 9. Los Angeles.

Cappannari, S. C.

1950 *The Concept of Property Among Shoshoneans*. Ph.D. Dissertation, Department of Anthropology, University of California, Berkeley.

1960 The Concept of Property Among Shoshoneans. In *Essays in the Science of Culture in Honor of Leslie A. White*, G. E. Cole and R. L. Carneiro, editors, pp. 133-144. New York: Thomas Crowell Company.

Carrico, R. L.

1982 *Phoenix Buttons from the San Diego Presidio Chapel Complex*. Report on file, San Diego Historical Society, San Diego.

Carroll, A. K., M. N. Zedeño and R. W. Stoffle

2004 Landscapes of the Ghost Dance: A Cartography of Numic Ritual. *Journal of Archaeological Method and Theory* 11(2):127-156.

Chalfant, W. A.

1933 *The Story of Inyo*. Bishop: Chalfant Press.

Clewlow, C. W., Jr.

1968 Surface Archaeology of the Black Rock Desert, Nevada. *University of California Archaeological Survey Reports* 73:1-94.

Clewlow, C. W., Jr., R. F. Heizer, and R. Berger

1970 An Assessment of the Radiocarbon Dates for the Rose Spring Site (CA-Iny-372), Inyo County, California. In Papers on Anthropology in the Great Basin, pp. 19-27. *University of California Archaeological Research Facility Contributions* 7, Berkeley.

Clewlow, C. W., Jr., S. Wallman, and T. Clewlow

1995 *Archaeological Test Excavations at Sites CA-INY-2845, CA-INY-2845, and CA-INY-2847, Inyo County, California*. Report on file with the Naval Air Weapons Station, China Lake, California.

Coombs, G. B. and R. S. Greenwood

1984 *A Cultural Resources Overview and Inventory Plan for the Naval Weapons Center, China Lake*. Report on file with the Naval Air Weapons Station, China Lake, California.

Coville, F. V.

1892 The Panamint Indians of California. *American Anthropologist* 5(4):351-361.

Cuevas, K. M.

2002 *Archaeological Investigations at The Long Canyon Village Site, CA-Ker-311, Kern County, California*. M.A. Thesis, Department of Anthropology, California State University, Bakersfield.

Davis, E. L

1978 The Ancient Californians: Rancholabrean Hunters of the Mojave Lakes Country. *Natural History Museum of Los Angeles County, Science Series* 29. Los Angeles.

Davis, W. B.

1938 Summer Activity of Mountain Sheep on Mount Washburn, Yellowstone, National Park, Wyoming. *Journal of Mammology* 19(1):88-94.

DeAtley, S. P. and F. Findlow

1984 *Exploring the Limits: Frontiers and Boundaries in Prehistory*. B.A.R. 223. Great Britain International Series.

Delacorte, M. G.

1985 The George T. Hunting Complex, Deep Springs Valley, California. *Journal of California and Great Basin Anthropology* 7:225-239.

1990 *The Prehistory of Deep Springs Valley, Eastern California: Adaptive Variation in the Western Great Basin*. Ph.D. Dissertation, Department of Anthropology, University of California, Davis.

1994 Late Prehistoric Resource Intensification in the Numic Heartland. Paper presented at the 24th Annual Great Basin Anthropological Conference, Elko, Nevada.

1995 Desert Side-notched Points as a Numic Population Marker in the West-Central Great Basin. Paper presented at the 29th Annual Meeting of the Society for California Archaeology, Eureka, California.

1999 *The Changing Role of Riverine Environments in the Prehistory of the Central-Western Great Basin: Data Recovery Excavations at Six Prehistoric Sites in Owens Valley, California*. Report on file California Department of Transportation, District 09, Bishop.

Delacorte, M. G., M.C. Hall and M. E. Basgall

1995 *Final Report on the Evaluation of Twelve Archaeological Sites in the Southern Owens Valley, Inyo County, California.* Report on file, California Department of Transportation, District 09, Bishop.

Delacorte, M. G. and K. R. McGuire

1993 *Report of Archaeological Test Evaluations at 23 Sites in Owens Valley, California.* Report on file, Contel of California, Inc, and the Department of the Interior, Bureau of Land Management, California Desert District, Ridgecrest.

Diebold, A. R., Jr.

1987 Linguistic Ways to Prehistory. In *Proto-Indo-European: The Archaeology of a Linguistic Problem: Studies in Honor of Marija Gimbutas*, S.N. Skomal and E. C. Polone, editors, pp. 19-71. Institute for the Study of Man, Washington, D.C.

Dietz, S.

1976 *Echa Tamal: A Study of Coast Miwok Acculturation.* M.A. thesis, Department of Anthropology, California State University, San Francisco.

Dillon, B. D.

1988 Southern Sierra Nevada Obsidian Hydration: Kern County Isabella Basin. In *Obsidian Dates, IV: A Compendium of the Obsidian Hydration Determinations Made at the U.C.L.A. Obsidian Hydration Laboratory*, C. W. Meighan and J. Scalise, editors, pp. 64-69. Monograph No. 29. Institute of Archaeology, University of California, Los Angeles.

2002 California PalaeoIndians: Lack of Evidence, or Evidence of a Lack? In Essays in California Archaeology: A Memorial to Franklin Fenenga, edited by W. J. Wallace and F. A. Riddell, pp. 25-54. *Contributions of the University of California Archaeological Research Facility* 60. Berkeley.

Drews, M. P. and R. G. Elston

1983 *An Archaeological Investigation of Drilling Locations and Power Plant Sites in the Coso Known Geothermal Area, China Lake Naval Weapons Center.* Report on file Naval Air Weapons Station, China Lake, California.

Driver, H. E.

1937 Cultural Element Distributions, VI: Southern Sierra Nevada. *University of California Anthropological Records* 1(2):53-154. Berkeley.

DuBois, C. A.

1939 The 1870 Ghost Dance. *University of California Anthropological Records* 3(1):1-151. Berkeley.

Dutcher, B. H.

1893 Piñon Gathering Among the Panamint Indians. *American Anthropologist* 6(4):377-380.

Eerkens, J. W. and J. H. King

2002 *Phase II Archaeological Investigations for the Sherwin Summit Rehabilitation Project, U.S. Highway 395, Inyo and Mono Counties, California.* Report on file, California Department of Transportation, District 09, Bishop.

Eerkens, J. W. and J. S. Rosenthal

2004 Are Obsidian Subsources Meaningful Units of Analysis?: Temporal and Spatial Patterning of Subsources in the Coso Volcanic Field, Southeastern California. *Journal of Archaeological Science* 31:21-29.

Ehret, C.

1976 Linguistic Evidence and its Correlation with Archaeology. *World Archaeology* 8:5-18.

1988 Language Change and the Material Cultural Correlates of Language and Ethnic Shifts. *Antiquity* 62:564-574.

Elston, R. G.

1994 How Will I Know You? Archeological Visibility of the Numic Spread in the Western Great Basin. In *Across the West: Human Population Movement and the Expansion of the Numa*, D. B. Madsen and D. Rhode, editors, pp. 150-151. Salt Lake City: University of Utah Press.

Elston, R.G. and C.D. Zeier

1984 The Sugarloaf Obsidian Quarry. *Naval Weapons Administrative Center, Publication 313*, China Lake, California.

Ericson, J. E.

1977 *Prehistoric Exchange Systems in California: The Results of Obsidian Dating and Tracing.* Ph.D. Dissertation, Department of Anthropology, University of California, Los Angeles.

1978a Obsidian Hydration Dating in California. *Society for California Archaeology, Occasional Papers in Method and Theory in California Archaeology*, No. 2, pp. 43-57.

1978b Date Lists, INY-372. In *Obsidian Dates II, A Compendium of the Obsidian Hydration Determinations Made at the U.C.L.A. Obsidian Hydration Laboratory*, C. W. Meighan and P.I. Vanderhoeven, editors, pp. 18-19. Monograph VI, Institute of Archaeology. University of California, Los Angeles.

Ericson, J. E. and M. D. Glascock

2004 Subsource Characterization: Obsidian Utilization of Subsources of the Coso Volcanic Field, Coso Junction, California, USA. *Geoarchaeology* 19(8):779-805.

Ericson, J. E. and C. W. Meighan

1984 Boundaries, Alliance, and Exchange in California. In *Exploring the Limits: Frontiers and Boundaries in Prehistory*, S. P. DeAtley and F. J. Findlow, editors, pp 143-152. B.A.R. 223. Great Britain International Series.

Eshleman, J. A.

2002 *Mitochondrial DNA and Prehistoric Population Movements in Western North America.* Ph.D. Dissertation, Department of Anthropology, University of California, Davis.

Essene, F. J.

1935 *Bishop and Big Pine Paiute Ethnographic Notes.* On file University Archives, Bancroft Library, University of California, Berkeley.

Farmer, M. F.

1937 An Obsidian Quarry near Coso Hot Springs. *Masterkey* 11:7-9.

Fenenga, F.

1977 *Cultural Resources Inventory, Hidden Dam – Hensley Lake.* Report on file, National Park Service, San Francisco.

Fiedel, S. J.

1999 Older Than We Thought: Implications for Corrected Dates for Paleoindians. *American Antiquity* 64(1):95-115.

Foster, M. K.

1996 Language and Cultural History of North America. In Handbook of North American Indians, Volume 17, *Languages*, pp. 64-115, I. Goddard, editor. Washington, D.C.: Smithsonian Institution Press.

Fowler, C. S.

1972 Some Ecological Clues to Proto-Numic Homelands. In Great Basin Cultural Ecology: A Symposium, D. D. Fowler, editor, pp. 105-122. *Desert Research Institute Publication in the Social Sciences* No. 8, Reno and Las Vegas.

1983 Some Lexical Clues to Uto-Aztecan Prehistory. *International Journal of American Linguistics* 49(3):224-257.

Fowler, D. D. and D. B. Madsen

1986 Prehistory of the Southeastern Area. In *Great Basin*, W. L. d'Azevedo, editor, pp. 173-182. Handbook of North American Indians, Volume 11, W. G. Sturtevant, General Editor. Washington D. C.: Smithsonian Institution.

Friedman, I. and R. Smith

1960 A New Dating Method Using Obsidian: Part I, The Development of the Technique. *American Antiquity* 25:476-493.

Garfinkel, A. P.

1976 *A Cultural Resource Management Plan for the Fossil Falls/Little Lake Locality.* Bakersfield District Office, Bureau of Land Management, Bakersfield.

1978 "Coso" Style Pictographs of the Southern Sierra Nevada. *Journal of California Anthropology* 5(1):94-101.

1982 The Identification of Prehistoric Aboriginal Groups through the Study of Rock Art. In Pictographs of the Coso Region, R. A. Schiffman, D. S. Whitley, A. P. Garfinkel, and S. B. Andrews, editors, pp. 67-78. *Bakersfield College Publications in Archaeology* No. 2. Bakersfield.

2003 Dating "Classic" Coso Style Sheep Petroglyphs in the Coso Range and El Paso Mountains: Implications for Regional Prehistory. *Society for California Archaeology Newsletter* 37(4):34-37

Garfinkel, A. P., J. D. Binning, E. Younkin, C. Skinner, T. Origer, R. Jackson and T. Carpenter

2004 The Little Lake Biface Cache, Inyo County, California. *Proceedings of the Society for California Archaeology* 17:87-101.

Garfinkel, A. P. and R. A. Cook

1979 Prehistoric Change in Central Eastern California: The Sherwin Grade Site. *California Department of Transportation Occasional Papers in Archaeology* 1.

1981 Radiocarbon Dating of Pinyon Nut Exploitation in Eastern California. *Journal of California and Great Basin Anthropology* 2(2): 283-285.

Garfinkel, A. P., L. S. Kobori, J. C. Bard, and R. J. Dezzani

1984 *Rockhouse Basin Data Recovery Program.* Report on file, United States Department of Agriculture, United States Forest Service, Sequoia National Forest, Porterville, California.

Garfinkel, A. P., G. Marcom, and R. A. Schiffman

2007 Culture Crisis and Rock Art Intensification: Numic Ghost Dance Paintings and Coso Representational Petroglyphs. In press, *American Indian Rock Art*, Volume 33, Don Christensen, editor. Tucson: American Rock Art Research Association.

Garfinkel, A. P. and J. K. Pringle

2004 Dating the Rock Drawings of the Coso Range: Projectile Point Petroglyphs. *American Indian Rock Art,* Volume 30, J. T. O'Connor, editor, pp. 1-14. Tucson: American Rock Art Research Association.

Garfinkel, A. P., R. A. Schiffman and K. R. McGuire

1980 Archaeological Investigations in the Southern Sierra Nevada: The Lamont Meadow and Morris Peak Segments of the Pacific Crest Trail. *Cultural Resources Publications, Archaeology.* Department of the Interior, Bureau of Land Management, Bakersfield District, Bakersfield.

Garfinkel, A. P., J. Wehausen, C. Epps and R. M. Yohe, II

2005 *Prehistoric Bighorn Depletion in the Coso Range of Eastern California, A Research Proposal.* Proposal on file, Naval Air Weapons Station, China Lake, California.

Garfinkel, A. P. and R. M. Yohe, II

2004 Antiquity and Function: Humboldt Basal-notched Bifaces in the Southwestern Great Basin. *Journal of California and Great Basin Anthropology* 24(1):103-125.

Gayton, A. H.

1948 Yokuts and Western Mono-Ethnography. *University of California Anthropological Records* 10(1-2):1-302. Berkeley.

Gehr, E. A.

1981 Obsidian Hydration Data from Various Tulare County Sites: Tulare River Indian Reservation. In *Obsidian Dates III: A Compendium of the Obsidian Hydration Determinations Made at the U.C.L.A. Obsidian Hydration Laboratory*, C. W. Meighan and G. S. Russell, editors, p. 35. University of California, Los Angeles.

1988 Mean Temperature Scaling of Source Specific Hydration Rates. In *Obsidian Dates IV: A Compendium of the Obsidian Hydration Determinations Made at the U.C.L.A. Obsidian Hydration Laboratory*, C. W. Meighan and J. L. Scalise, editors, pp. 19-26. University of California, Los Angeles.

Geist, V.

1967 A Consequence of Togetherness. *Natural History Magazine* 76(8):24-30.

1971 *Mountain Sheep: A Study in Behavior and Evolution.* Chicago and London: University of Chicago Press.

Gibson, R. O.

1975 The Beads of *Humaliwo*. *Journal of California and Great Basin Anthropology* 2:110-119.

Gifford, E. W.

1936 Northeastern and Western Yavapai. *University of California Publication in American Archaeology and Ethnology* 34(4). Berkeley.

Gilreath, A. J.

1992 *Draft Report of Archaeological Test Excavations at Three Sites in the Fossil Falls Archaeological District – CA-INY-1634, INY-1635, and INY-1662.* Report on file, Ridgecrest Resource Area, Bureau of Land Management, Ridgecrest.

1997 *Cultural Resources Inventory in Support of the Coso Rock Art District/Landmark Boundary Study* Volume 1. Report on file, Naval Air Weapons Station, China Lake, California.

1999 *Chronological Assessment of the Coso Rock Art Landmark – An Obsidian Hydration Analysis.* Report on file, Naval Air Weapons Station, China Lake, California.

2000 *Archaeological and Osteological Report in Support of the Native American Graves Protection And Repatriation Act, Naval Air Weapons Station, China Lake.* Report on file, Naval Air Weapons Station, China Lake, California.

2003 Age and Function of Rock Art. In *Archaeological Testing of Fourteen Prehistoric Sites within the Coso Target Range at Naval Air Weapons Station, China Lake, California* by W. Hildebrandt and A. Ruby, pp. 209-214. Report on file, Naval Air Weapons Station, China Lake, California.

Gilreath, A. J. and W.R. Hildebrandt

1997 Prehistoric Use of the Coso Volcanic Field. *Contributions of the University of California Archaeological Research Facility* No. 56. University of California, Berkeley

Gilreath, A.J. and K. Holanda

2000 *By the Lake By the Mountains: Archaeological Investigations at CA-INY-4554 and CA-INY-1428.* Report on file, California Department of Transportation, District 9, Bishop.

Glennan, W. S.

1971 Concave-based Lanceolate Fluted Projectile Points from California. *Masterkey* 45(1):27-32.

Golla, V.

2000 Language History and Communicative Strategies in Aboriginal California and Oregon. In *Languages of the North Pacific Rim*, Volume 5, O. Miyaoka and M. Oshima, editors, pp. 43-64. Suita, Japan: Faculty of Informatics, Osaka Gakuin University.

Goss, J. A.

1964 Cultural Development in the Great Basin. *Utah Archaeology* 10(4):4-12.

1965 Ute Linguistics and Anasazi Abandonment of the Four Corners Area. In Contributions of the Wetherill Mesa Archaeological Project assembled by D. Osborne. *Memoirs of the Society for American Archaeology* No. 199, pp. 73-81. Washington, D.C.

1968 Culture-Historical Inference from Utaztecan Linguistic Evidence. In Utaztecan Prehistory, E. H. Swanson, Jr., editor, pp. 10-42. *Occasional Papers of the Idaho State University Museum* No. 22. Pocatello.

1977 Linguistic Tools for the Great Basin Prehistorian. In Models and Great Basin Prehistory: A Symposium, D. D. Fowler, editor, pp. 48-70. *Desert Research Institute Publication in the Social Sciences* No. 12, Reno and Las Vegas.

Grant, C., J.W. Baird, and K. Pringle

1968 Rock Drawings of the Coso Range, Inyo County, California: An Ancient Sheep-hunting Cult Pictured in Desert Rock Carvings. *Maturango Museum Publication* 4. China Lake, California.

Graumlich, L. J.

1993 A 1000-year Record of Temperature and Precipitation in the Sierra Nevada. *Quaternary Research* 39:249-255.

Graybill, D. A., M. R. Rose and F. L. Nials

1994 Tree-rings and Climate: Implications for Great Basin Paleoenvironmental Studies. In *Proceedings of the International High-level Radioactive Waste Management Conference and Exposition*, Las Vegas, Nevada.

Grayson, D. K.

1994 Chronology, Glottochronology, and Numic Expansion. In *Across the West: Human Population Movement and the Expansion of the Numa*, D. B. Madsen and D. Rhode, editors, pp. 20-23. Salt Lake City: University of Utah Press.

Griset, S.

1981 Pottery Analysis. In *Archaeological Investigations in the Southern Sierra Nevada: The Kennedy Meadows and Rockhouse Basin Segments of the Pacific Crest Trail* by K. R McGuire, pp. 186-197. Report on file, United States Department of Agriculture, United States Forest Service, Sequoia National Forest, Porterville, California.

Grosscup, G. L.

1977 Notes on Boundaries and Culture of the Panamint Shoshone and Owens Valley Paiute. *University of California Archaeological Research Facility Contributions* 35:109-150. Berkeley.

Gudschinsky, S.

1956 The ABCs of Lexicostatistics (Glottochronology). *Word* 12:175-210.

Hale, K.

1958 Internal Diversity in Uto-Aztecan. *International Journal of American Linguistics* 24(2):104-107. Baltimore.

Halford, F. K.

1998 *Archaeology and Environment on the Dry Lakes Plateau, Bodie Hills, California: Hunter-Gatherer Coping Strategies for Holocene Environmental Variability.* M.A. Thesis, Department of Anthropology, University of Nevada, Reno.

2001 New Evidence for Early Holocene Acquisition and Production of Bodie Hills Obsidian. *Society for California Archaeology Newsletter* 35(1):32-37.

Hall, M. C.

1983 *Late Holocene Hunter-Gatherers and Volcanism in the Long Valley-Mono Basin Region: Prehistoric Culture Change in the Eastern Sierra Nevada.* Ph.D. Dissertation, Department of Anthropology, University of California, Riverside.

Hall, M. C. and R. J. Jackson

1989 Obsidian Hydration Rates in California. In Current Directions in California Obsidian Studies, R. E. Hughes, editor, pp. 31-58. *Contributions of the University of California Archaeological Research Facility* 48. University of California, Berkeley.

Harrington, J. P.

1934 *Ethnographic Field Notes, Tubatulabal.* J. P. Harrington Collection, National Anthropological Archives, Smithsonian Institution, Washington. [Microfilm edition, Millwood, New York: Kraus International Publications.]

Harrington, M. R.

1933 Gypsum Cave, Nevada. *Southwest Museum Papers* No. 8. Los Angeles.

1957 A Pinto Site at Little Lake, California. *Southwest Museum Papers* No. 17. Los Angeles.

Haury, E.

1945 Painted Cave, Northeastern Arizona. *Amerind Foundation Paper* No. 3.

Hayes, M. G., J. B. Coltrain, and D. H. O'Rourke

2002 Paleogenetic Assessment of Human Migration and Population Replacement in North American Arctic Prehistory. Paper presented at the ARCUS 14th Annual Meeting and Arctic Forum, Arlington, Virginia.

Heizer, R. F. and M. A. Baumhoff

1961 The Archaeology of Wagon Jack Shelter. *University of California Anthropological Records* 20(4):119-138. Berkeley.

1962 *Prehistoric Rock Art of Nevada and Eastern California.* Berkeley: University of California Press.

Heizer, R. F., M. A. Baumhoff, and C. W. Clewlow, Jr.

1968 Archaeology of South Fork Shelter (NV-EL-11), Elko County, Nevada. *University of California Archaeological Survey Reports* 71:1-58. Berkeley.

Heizer, R. F. and C. W. Clewlow, Jr.

1968 Projectile Points from NV-Ch-15, Churchill County, Nevada. *University of California Archaeological Survey Reports* 71:59-88. Berkeley.

1973 *Prehistoric Rock Art of California*. Ramona: Ballena Press.

Heizer, R. F. and T. R. Hester

1974 Two Petroglyph Sites in Lincoln County, Nevada. *University of California Contributions, Archaeological Research Facility* 20:1-52.

1978 Great Basin. In *Chronologies in New World Archaeology*, R. E. Taylor and C. W. Meighan, editors, pp. 147-199. New York: Academic Press.

Hester, T. R.

1973 Chronological Ordering of Great Basin Prehistory. *University of California Archaeological Research Facility Contributions* 17. Berkeley.

Hickman, J. C. (editor)

1993 *The Jepson Manual: Higher Plants of California*. Berkeley: University of California Press.

Hildebrandt, W. R. and K. R. McGuire

2003 The Ascendance of Hunting During the California Middle Archaic: An Evolutionary Perspective. *American Antiquity* 67(2):231-256.

Hildebrandt, W. R. and A. Ruby

1999 *Prehistoric Pinyon Exploitation in the Southwestern Great Basin: A View from the Coso Range*. Manuscript on file, Far Western Anthropological Research Group, Davis, California.

2003 Archaeological Testing of Fourteen Prehistoric Sites within the Coso Target Range at Naval Air Weapons Station, China Lake, California. Report on file, Naval Air Weapons Station, China Lake, California.

Hill, J. H.

2002 *Language Spread Among Hunter-Gatherers*. Website accessed November 2004, *http://crclc.anu.edu.au/arcling2/Hill.htn*.

Hill, M.

2005 *Geology of the Sierra Nevada* (California Natural History Guides, 80). Revised Edition. Berkeley: University of California Press.

Hillebrand, T. S.

1972 *The Archaeology of the Coso Locality of the Northern Mojave Region of California*. Ph.D. Dissertation, Department of Anthropology, University of California, Santa Barbara.

Hittman, M.

1973 The 1870s Ghost Dance at the Walker River Reservation: A Reconstruction. *Ethnohistory* 20(3):247-278.

1997 *Wovoka and the Ghost Dance*, Expanded Edition. Lincoln: University of Nebraska Press.

Hodder, I.

1982 *Symbols in Action: Ethnoarchaeological Studies of Material Culture*. Cambridge: Cambridge University of Press.

Hughes, R. E.

1992 California Archaeology and Linguistic Prehistory. *Journal of Anthropological Research* 48:317-338.

1994 Methodological Observations on Great Basin Prehistory. In *Across the West: Human Population Movement and the Expansion of the Numa*, D. B. Madsen and D. Rhode, editors, pp. 67-70. Salt Lake City: University of Utah Press.

Hull, K. L. and M. J. Moratto

1999 Archaeological Synthesis and Research Design, Yosemite National Park, California, edited by Michael J. Moratto. USDI National Park Service, *Yosemite Research Center Publications in Anthropology* 21. Yosemite National Park.

Hultkrantz, A.

1986 Mythology and Religious Concepts. In Handbook of North American Indians, Volume 11, *Great Basin*, W. L. d'Azevedo, editor, pp. 630-640. Washington, D.C.: Smithsonian Institution.

Hymes, D. H.

1960 Lexicostatistics So Far. *Current Anthropology* 1:3-44.

Irwin, C. N. (editor)

1980 The Shoshoni Indians of Inyo County, California: The Kerr Manuscript. *Ballena Press Publications in Archaeology, Ethnology and History* No.15. Socorro, New Mexico.

Jackson, R. J.

1984 Obsidian Hydration: Applications in the Western Great Basin in Obsidian Studies Studies in the Great Basin, R. E. Hughes, editor, pp. 173-192. *University of California Archaeological Research Facility Contributions* No. 45.

1985 *An Archaeological Survey of the Wet, Antelope, Railroad, and Ford Timber Sale Compartments in the Inyo National Forest.* Report on file, United States Forest Service, Inyo National Forest, Bishop.

Jennings, S. A. and D. L. Elliot-Fisk

1993 Packrat Midden Evidence of Late Quaternary Vegetation Change in the White Mountains, California-Nevada. *Quaternary Research* 39:214-221.

Jones, T. L., G. M. Brown, L. M. Raab, J. L. McVickar, W. G. Spaulding, D. J. Kennett, A. York, and P. L. Walker

2004 Environmental Imperatives Reconsidered: Demographic Crises in Western North America During the Medieval Climatic Anomaly. In *Prehistoric California: Archaeology and the Myth of Paradise*, L. M. Raab and T. L. Jones, editors, pp. 12-32. Salt Lake City: University of Utah Press.

Jones, T. L. and G. Waugh

1994 Central California Coastal Prehistory: A View from Little Pico Creek. *Perspective in California Archaeology*, Volume 3. Institute of Archaeology, University of California, Los Angeles.

Justice, N. D.

2002 *Stone Age Spear and Arrow Points of California and the Great Basin*. Bloomington: Indiana University Press.

Kaestle, F. A.

1998 *Molecular Evidence for Prehistoric Native American Population Movement: The Numic Expansion*. Ph.D. Dissertation, Department of Anthropology, University of California, Davis.

Kaestle, F. A. and K. A. Horsburgh

2002 Ancient DNA in Anthropology: Methods, Applications and Ethics. *Yearbook of Physical Anthropology* 45:92-130.

Kaestle, F. A. and D. G. Smith

2001 Ancient Mitochondrial DNA Evidence for Prehistoric Population Movement: The Numic Expansion. *American Journal of Physical Anthropology* 115:1-12.

Kelly, I. T. and C. S. Fowler

1986 Southern Paiute. In Handbook of North American Indians, Volume 11, *Great Basin*, W. L. d'Azevedo, editor, pp. 525-557. Washington, D.C.: Smithsonian Institution.

Kemp, B. M., D. G. Smith and W. J. Nelson.

2004 DNA Analysis of Ancient Coprolites from Fish Slough Cave. Poster presented at the Great Basin Anthropological Conference.

Kidd, K. and M. A. Kidd

1970 A Classification System for Glass Beads for the Use of Field Archaeologists. Canadian Historic Sites: *Occasional Papers in Archaeology and History* 1:45-89.

King, C. D.

1990 *The Evolution of Chumash Society: A Comparative Study of Artifacts Used in Social Systems Maintenance in the Santa Barbara Channel Region Before A.D. 1804.* New York: Garland Publishing Company.

King, J.

2000 Re-examining the Coso Obsidian Hydration Rates. Paper presented at the Society for California Archaeology Meetings.

King, T.

1968 The Archaeology of the Schwabacher Site, 4-Made-117. *San Francisco State College, Anthropology Museum, Occasional Papers* 4(2):1-135.

Kroeber, A. L.

1907 Shoshonean Dialects of California. *University of California Publications in American Archaeology and Ethnography* No. 4(3). Berkeley.

1925 *Handbook of the Indians of California.* Bureau of American Ethnology Bulletin 78. Washington, D.C.

1955 Linguistic Time Depth Results So Far and Their Meaning. *International Journal of American Linguistics* 21(2):91-104.

1959 Ethnographic Interpretations 7-11. *University of California Publications in American Archaeology and Ethnology* 47(3):259-281. Berkeley.

LaMarche, V. C., Jr.

1973 Holocene Climatic Variations Inferred from Treeline Fluctuations in the White Mountains of California. *Quaternary Research* 3(4):632-660.

1974 Paleoclimatic Inferences from Long Tree-Ring Records. *Science* 183:1043-1048.

Lamb, S.

1958 Linguistic Prehistory of the Great Basin. *International Journal of American Linguistics* 24(2):95-100.

Lanning, E. P.

1963 Archaeology of the Rose Spring Site (INY-372). *University of California Publications in American Archaeology and Ethnology* 49(3):237-336. Berkeley.

Layton, R.

1986 Political and Territorial Structures among Hunter-Gatherers. *Man* 21:18-33.

Layton, T. N.

1985 Invaders from the South? Archaeological Discontinuities in the Northwestern Great Basin. *Journal of California and Great Basin Anthropology* 7(2):183-201.

Lee, G.

1991 *Rock Art Sites at Tomo-Kahni, Kern County, California.* Report on file, California Department of Parks and Recreation, Sacramento.

Lee, G. and W. D. Hyder

1991 Prehistoric Rock Art as an Indicator of Cultural Interaction and Tribal Boundaries in South-central California. *Journal of California and Great Basin Anthropology* 13(1):15-28.

Leopold, A. S.

1951 The Jawbone Deer Herd. *California Fish and Game Bulletin* No. 4.

Lindstrom, R. W.

1996 Linguistic Expansion, Ethnogenetic Models and Culture Change in the Bronze Age of the Southern Zaural. Paper presented at the 95th Annual Meeting of the American Anthropological Association, Atlanta, Georgia.

Longhurst, W. M., A. S. Leopold, and R. F. Dasmann

1952 A Survey of California Deer Herds: Their Ranges and Management Problems. *State of California Fish and Game Bulletin* No. 6.

McCarthy, H. and L. Johnson

2002 Appendix A: Ethnography and Ethnohistory: Owens Valley Paiute. In *Final Report on Phase II Investigations at 26 Archaeological Sites for the Aberdeen-Blackrock Four Lane Project on Highway 395, Inyo County, California* by D. Zeanah and A. Leigh. Report on file, California Department of Transportation, Fresno.

McClean, D. D.

1944 The Prong-horned Antelope in California. *California Fish and Game* 30:221-241

McGrath, R. D.

1987 *Gunfighters, Highwaymen and Vigilantes: Violence on the Frontier.* Berkeley: University of California Press.

McGuire, K. R.

1981 *Archaeological Investigations in the Southern Sierra Nevada: The Kennedy Meadows/Rockhouse Basin Segment of the Pacific Crest Trail.* Report on file, United States Department of Agriculture, United States Forest Service, Sequoia National Forest, Porterville, California.

1983 *Archaeological Investigations in the Scodie Mountains of the Southern Sierra Nevada: CA-KER-1286.* Report on file, United States Department of Agriculture, United States Forest Service, Sequoia National Forest, Porterville, California.

1989 Incised Stones. In Prehistory of the Sacramento River Canyon, California by M. E. Basgall and W. R. Hildebrandt, Appendix D, pp. 1-43. *Center for Archaeological Research at Davis, Publication* Number 9.

McGuire, K. R. and A. P. Garfinkel

1976 Comment on "The Development of Pinyon Exploitation in Central Eastern California." *Journal of California Anthropology* 3(2):83-85.

1980 Archaeological Investigations in the Southern Sierra Nevada: The Bear Mountain Segment of the Pacific Crest Trail. *Cultural Resources Publications, Archaeology.* Department of the Interior, Bureau of Land Management, Bakersfield District, Bakersfield.

McGuire, K. R., A. P. Garfinkel, and M.E. Basgall

1982 *Archaeological Investigations in the El Paso Mountains of the Western Mojave Desert: The Bickel and Last Chance Sites, CA-Ker-250, -261.* Report on file, Bureau of Land Management, Ridgecrest.

Madsen, D. B.

1975 Dating Paiute-Shoshone Expansion in the Great Basin. *American Antiquity* 40(1):82-86.

1981 The Emperor's New Clothes. *American Antiquity* 46(3):637-640.

1986 Great Basin Nuts: A Short Treatise on the Distribution, Productivity and Prehistoric Use of Pinyon. In *Anthropology of the Desert West: Essays in Honor of Jesse D. Jennings*, D. J. Condie and D. D. Fowler, editors, pp. 21-42. Salt Lake City: University of Utah Press.

1994 Mesa Verde and Sleeping Ute Mountain: The Geographical and Chronological Dimensions of the Numic Expansion. In *Across the West: Human Population Movement and the Expansion of the Numa*, D. B. Madsen and D. Rhode, editors, pp. 24-31. Salt Lake City: University of Utah Press.

Madsen, D. B. and D. Rhode (editors)

1994 *Across the West: Human Population Movement and the Expansion of the Numa.* Salt Lake City: University of Utah Press.

Marcom, G.

2002 *The Pictographs of Death Valley, An Inventory and Assessment.* Report on file, Death Valley National Park, Death Valley, California.

Markos, J. A., L. A. Walsh, L. Hause, T. D. Burke, and C. L. Rogers

1995 *Cool, Cool Water: Archaeology of the Pa Doya Site (CA-Iny-1991), Southern Owens Valley, Inyo County, California.* Report submitted to Inyo County Planning Department, Independence. Report on file, Eastern Information Center, University of California, Riverside.

Marwitt, J. P.

1986 Fremont Cultures. In *Great Basin*, W. L. d'Azevedo, editor, pp. 161-172. Handbook of North American Indians, Volume 11. Washington, D.C.: Smithsonian Institution.

May, R. V.

1980 Ceramic Artifacts. In Archaeological Investigations in the Southern Sierra Nevada: The Lamont Meadow and Morris Peak Segments of the Pacific Crest Trail by A. P. Garfinkel, R. A. Schiffman and K. R. McGuire, pp. 276-279. *Cultural Resources Publications, Archaeology*. Department of the Interior, Bureau of Land Management, Bakersfield District, Bakersfield.

1981 An Analysis of Ceramics from the Scodie Mountains Segment of the Pacific Crest Trail, Sequoia National Forest. In *Archaeological Investigations for the Scodie Mountains Segment of the Pacific Crest Trail, Kern County, California* by R. D. Ambro, M. K. Crist, B. J. Peck, and D. M. Varner. Appendix E. Report on file, United States Department of Agriculture, United States Forest Service, Sequoia National Forest, Porterville, California.

Mehringer, P. J., Jr.

1977 Great Basin Late Quaternary Environments and Chronology. In Models and Great Basin Prehistory, D. D. Fowler, editor, pp. 113-167. *Desert Research Institute Publications in the Social Sciences* 12. Reno.

1986 Prehistoric Environments. In *Great Basin*, W. L. d'Azevedo, editor, pp. 31-50. Handbook of North American Indians, Volume 11. Washington, D.C.: Smithsonian Institution.

Mehringer, P. J., Jr. and J. C. Sheppard

1978 Holocene History of Little Lake, Mojave Desert, California. In The Ancient Californians: Rancholabrean Hunters of the Mojave Lakes Country, edited by E. L. Davis, pp. 153-166. *Natural History Museum of Los Angeles County, Science Series* 29.

Meighan, C. W.

1955 Notes on the Archaeology of Mono County, California. In Papers on California Archaeology, pp. 6-28. *University of California Archaeological Survey Reports* 28. Berkeley.

1978 Empirical Determination of Obsidian Hydration Rates from Archaeological Evidence. In *Advances in Obsidian Glass Studies: Archaeological and Geochemical Perspectives*, R. E. Taylor, editor. Noyes Press, Park Ridge, New Jersey.

1981 The Little Lake Site, Pinto Points, and Obsidian Dating in the Great Basin. *Journal of California and Great Basin Anthropology* 3(2):200-214.

Merriam, C. H.

1904 Distribution of Indian Tribes in the Southern Sierra and Adjacent Parts of the San Joaquin Valley, California. *Science* 19(494):912-917.

[1937-1938] [*Notes on Tubatulabal Ethnogeography and Vocabulary.*] (Manuscript in C. H. Merriam Collection, Anthropology Museum, University of California, Berkeley.)

Miller, W. J.

1966 Anthropological Linguistics in the Great Basin. In The Current Status of Anthropological Research in the Great Basin,W. L. D'Azevedo et al., editors, pp. 75-112. *University of Nevada, Desert Research Institute Social Sciences and Humanities Publications* No. 1. Reno.

1970 Western Shoshoni Dialects. In *Languages and Cultures of Western North American: Essays in Honor of Sven Liljeblad* E. H. Swanson, Jr., editor, pp. 17-36. Pocatello: Idaho State University Press.

1972 Comments on Cultural Ecology Papers. In Great Basin Cultural Ecology: A Symposium, D. D. Fowler, editor, pp. 150-160. *Desert Research Institute Publications in Social Sciences* No. 8. Reno and Los Angeles.

1984 The Classification of Uto-Aztecan Languages Based on the Lexical Evidence. *International Journal of American Linguistics* 50(1):1-24.

1986 Numic Languages. In Handbook of North American Indians, Volume 11, *Great Basin*, W. L. d'Azevedo, editor, pp. 98-107. Washington D. C.: Smithsonian Institution.

Miller, W. J., J. L. Tanner, and L. P. Foley

1971 A Lexicostatistical Study of Shoshoni Dialects. *Anthropological Linguistics* 13(4):145-164.

Milliken, R.

1999 Shell, Stone and Bone Beads in the Owens Valley. In *The Changing Role of Riverine Environments in the Prehistory of the Central-Western Great Basin: Data Recovery Excavations at Six Prehistoric Sites in Owens Valley, California* by M. G. Delacorte, pp. 65-108. Report on file, California Department of Transportation, District 9, Bishop.

Mooney, J.

1973 *The Ghost Dance Religion and Wounded Knee*. New York: Dover Publications.

Moratto, M. J.

1972 *A Study of the Prehistory of the Southern Sierra Nevada Foothills, California*. Ph.D. Dissertation, Department of Anthropology, University of Oregon.

1984 *California Archaeology*. Orlando: Academic Press.

1988 *Archaeological Excavations at Site CA-FRE-1671 Fresno County, California: Final Report*. Report on file, California Department of Transportation, District 6, Fresno.

Moratto M., T. F. King and W. B. Woolfenden

1978 Archaeology and California's Climate. *Journal of California Anthropology* 5(2):147-161.

Morwood, M. J.

1992 Introductory Essay on Ethnography and the Archaeological Study of Rock Art. In Rock Art and Ethnography, M. J. Morwood and D. Hobbs, editors, pp. 1-6. *Occasional AURA Publication* 5. Australian Rock Art Research Association, Melbourne.

Moyle, P. B.

2002 *Inland Fishes of California*. Berkeley: University of California Press.

Muir, J.

1901 *The Mountains of California*. New York: The Century Company.

Muto, G. R., P. J. Mehringer, and C. N. Warren.

1976 A Technological Analysis of Projectile Points from a Burial, Amargosa Desert, Nevada. *The Kiva* 41(3-4):267-276.

Nash, M.

1996 The Core Elements of Ethnicity. In *Ethnicity*, J. Hutchison and A. D. Smith, editors. Oxford: Oxford Press.

Nelson, E. W., Jr.

1891 The Panamint and Saline Valley (California) Indians. *American Anthropologist* 4(4):371-372.

1922 Lower California and Its Natural Resources. Washington, D.C.: *Memoirs of the National Academy of Sciences* 16:132-133.

Nelson, W.

1999 *A Paleodietary Approach to Late Prehistoric Hunter-gatherer Settlement-subsistence Change in Northern Owens Valley, Eastern California: the Fish Slough Cave Example*. Ph.D. Dissertation. Department of Anthropology, University of California, Davis.

Nichols, J.

1992 *Linguistic Diversity in Space and Time*. Chicago, Illinois: University of Chicago Press.

Nissen, K. M.

1974 A New Style of Rock Art in the Great Basin. The Stillwater Faceted Style. *Nevada Archaeological Survey* 8(2):10-11.

1982 *Images from the Past: An Analysis of Six Great Basin Petroglyph Sites*. Ph.D. Dissertation, Department of Anthropology, University of California, Berkeley.

1995 Pray for Signs? Petroglyph Research in the Western Great Basin, North America. In Rock Art Studies in the Americas, Papers from the Darwin Rock Art Congress, J. Steinberg, editor. *Oxbow Monograph* 45:67-75. Oxford, Great Britain.

O'Connell, J. F.

1967 Elko Eared/Elko Corner-notched Projectile Points as Time Markers in the Great Basin. *University of California Archaeological Survey Reports* 70:129-140. Berkeley.

Pearson, J. L.

1995 *Prehistoric Occupation at Little Lake, Inyo County, California: A Definitive Chronology*. M.A. Thesis, Department of Anthropology, Loyola University of Los Angeles.

2002 *Shamanism and the Ancient Mind, A Cognitive Approach to Archaeology.* Walnut Creek, California: Altamira Press.

Pendleton, L. S.

1979 *Lithic Technology in Early Nevada Assemblages.* M.A. Thesis, Department of Anthropology, California State University, Long Beach.

Peterson, J. T.

1978 The Ecology of Social Boundaries: Agta Foragers of the Phillipines. *Illinois Studies in Anthropology* No. 11. Urbana: University of Illinois Press.

Pippin, L. C.

1977 Bighorn Sheep and Great Basin Prehistory. *Nevada State Museum Anthropological Papers* Number 17:332-358.

Polanich, J. K.

2000 Basketry. In *Archaeological and Osteological Report in Support of the Native American Graves Protection and Repatriation Act, Naval Air Weapons Station, China Lake* by A. J. Gilreath, pp. 31-35. Report on file, Naval Air Weapons Station, China Lake, California.

Powell, J. W.

1891 Linguistic Families of North America. In Seventh *Annual Report of the Bureau of American Ethnology* for the Years 1885-1886, pp. 1-142. Washington, D.C.

Powers, B.

1971 *South Fork Country.* Tucson: Westernlore Press.

1974 *North Fork Country.* Tucson: Westernlore Press

1981 *Indian Country of the Tubatulabal.* Tucson: Westernlore Press.

Powers, S.

1877 Tribes of California. *Contributions to North American Ethnology* 3. Washington: United States Geographical and Geological Survey of the Rocky Mountain Region.

Quinlan, A. R. and A.Woody

2003 Marks of Distinction: Rock Art and Ethnic Identification in the Great Basin. *American Antiquity* 68(2):372-390.

Rector, C. H.

1985 Rock Art as Hunting Magic: Anthropological Fact or Fiction? *Rock Art Papers* 2:127-132.

Renfrew, C.

1987 *Archaeology and Language: The Puzzle of Indo-European Origins.* London: Jonathan Cape.

Renfrew, C., A. McMahon, and L. Trask

2000 *Time Depth in Historical Linguistics*, Volumes 1 and 2. Cambridge, England: Oxbow Books

Reynolds, L. A.

1996 *In the Dwelling Place of a Great Spirit: The Prehistory of the Piñon-Juniper Woodland of the Inyo-White Mountain Region.* Ph.D. Dissertation, Department of Anthropology, University of Nevada, Reno.

Rhode, D.

1980a Environmental Background and Ethnobotany. In Archaeological Investigations in the Southern Sierra Nevada: The Lamont Meadow and Morris Peak Segments of the Pacific Crest Trail by A. P. Garfinkel, R. A. Schiffman and K. R. McGuire, editors, pp. 16-37. *Cultural Resources Publications, Archaeology.* Department of the Interior, Bureau of Land Management, Bakersfield District, Bakersfield.

1980b Environmental Context. In Archaeological Investigations in the Southern Sierra Nevada: The Bear Mountain Segment of the Pacific Crest Trail, K. R. McGuire and A. P. Garfinkel, editors, pp. 6-14. *Cultural Resources Publications, Archaeology.* Department of the Interior, Bureau of Land Management, Bakersfield District, Bakersfield.

1980c Plant Macrofossils. In Archaeological Investigations in the Southern Sierra Nevada: The Lamont Meadow and Morris Peak Segments of the Pacific Crest Trail, A. P. Garfinkel, R. A. Schiffman and K. R. McGuire, editors, pp. 317-326.

Cultural Resources Publications, Archaeology. Department of the Interior, Bureau of Land Management, Bakersfield District, Bakersfield.

1980d Charred Plant Macrofossils. In Archaeological Investigations in the Southern Sierra Nevada: The Bear Mountain Segment of the Pacific Crest Trail, K. R. McGuire and A. P. Garfinkel, editors, p. 301. *Cultural Resources Publications, Archaeology*. Department of the Interior, Bureau of Land Management, Bakersfield District, Bakersfield.

Rhode, D. and D. B. Madsen

1998 Pine Nut Use in the Early Holocene and Beyond: The Danger Cave Archaeobotanical Record. *Journal of Archaeological Science* 25:1199-1210.

Riddell, F. A. and W. H. Olsen

1969 An Early Man Site in the San Joaquin Valley, California. *American Antiquity* 61(1):136-148.

Riddell, H. S., Jr.

1951 The Archaeology of a Paiute Village Site in Owens Valley. *University of California Archaeological Survey Reports* 12(15):14-28. Berkeley.

Riddell, H. S., Jr. and F. A. Riddell

1956 The Current Status of Archaeological Investigations in Owens Valley, California. *University of California Archaeological Survey Reports* 33(38):28-33. Berkeley.

Ritter, E. W.

1994 Scratched Rock Art Complexes in the Desert West: Symbols for Socio-Religious Communication. In New Light on Old Art, Recent Advances in Hunter-Gatherer Rock Art Research, D. S. Whitley and L. L. Loendorf, editors, pp. 51-66. *Institute of Archaeology Monograph* 36, University of California, Los Angeles.

2004 Paradigms and Petroglyphs of the Dry Lakes Plateau, Mono County, California. in *American Indian Rock Art*, Volume 30, edited by J. T. O'Connor, pp. 19-42. Tucson: American Rock Art Research Association.

Ritter, E., R. Brook and N. Farrel

1982. The Rock Art of Panamint City, Inyo County, California. In Pictographs of the Coso Region: Analysis and Interpretations of the Coso Painted Style, R. A. Schiffman, D. S. Whitley, A. P. Garfinkel, and S. B. Andrews, editors, pp. 5-21. *Bakersfield College Publications in Archaeology*, Number 2.

Roberts Jr., J. M., Moore, C. C. Romney and A. K. Romney

1995 Predicting Similarity in Material Culture Among the New Guinea Villages from Propinquity and Language. *Current Anthropology* 36(5):769-775.

Rogers, A. K.

2004 *Prehistoric Occupation Sequences in the El Paso Mountains, Kern County, California: An Analysis of Chronological Marker Artifacts from the Ron Henry Collection, Maturango Museum*. Report on file, Bureau of Land Management, Ridgecrest.

Rosenthal, J. S., K. L. Carpenter and D. C. Young, Jr.

2001 *Archaeological Survey of Target Area Buffer Zones in the Airport Lake, Baker, and George Ranges, Naval Air Weapons Station, China Lake, Inyo and Kern, Counties, California*. Report on file, Southwest Division, Naval Facilities Engineering Command, San Diego, California.

Ross, L. A.

1991 *Glass Beads from the 1977 University of Redlands Archaeological Excavations at the Southern Tip of the Yucaipa Rancheria Site (CA-SBR-1000/H, Yucaipa, San Bernardino County, California*. Report on file, Far Western Anthropological Research Group, Davis.

Ruhstaller, T.

1980 Rock Ring Features. In Archaeological Investigations in the Southern Sierra Nevada: The Bear Mountain Segment of the Pacific Crest Trail. *Cultural Resources Publications, Archaeology*. K. R. McGuire and A. P. Garfinkel, editors, pp. 200-204. Department of the Interior, Bureau of Land Management, Bakersfield District, Bakersfield.

Sackett, J. R.

1985 Style and Ethnicity in the Kalahari: A Reply to Weissner. *American Antiquity* 50(1):154-159.

1986 Style, Function, and Assemblage Variability: A Reply to Binford. *American Antiquity* 51(3):628-634.

Salzman, S.

1977 *The Valley of the South Fork of the Kern River: Cultural Resource Management Proposals for the Long Canyon Village Site.* Bureau of Land Management, Bakersfield, California.

Sapir, E.

1916 Time Perspective in Aboriginal American Culture: A Study in Method. *Anthropological Series* 13, Canada Department of Mines, Geological Survey Memoir 90, Ottawa.

1949 Time Perspective in Aboriginal American Culture: A Study in Method. In *Selected Writings of Edward Sapir*, D. G. Mandelbaum, editor, pp. 389-462. University of California Press.

Sawyer, J. O. and T. Keeler-Wolf

1995 *Manual of California Vegetation.* Sacramento: California Native Plant Society.

2006 *Manual of California Vegetation*, 2nd Edition. Sacramento: California Native Plant Society.

Schaafsma, P.

1986 Rock Art. In Handbook of North American Indians, Volume 11, *Great Basin*, W. L. d'Azevedo, editor, pp. 215-226. Washington, D. C.: Smithsonian Institution.

Schiffman, R. A.

1980 *Archaeological Investigations of a Tubatulabal Indian Hamlet, Sequoia National Forest, Kern County, California.* Report on file, United States Department of Agriculture, United States Forest Service, Sequoia National Forest, Porterville, California.

Schiffman, R. A. and S. B. Andrews

1979 The Slippery Rock Petroglyph Site: Analysis and Interpretation of Ker-25. In *American Indian Rock Art* Volume VI, F. G. Bock, editor, pp. 141-153. American Rock Art Research Association, El Toro, California.

1982 Pictographs of the Ghost Dance Movement of 1870 and 1890. In Pictographs of the Coso Region: Analysis and Interpretations of the Coso Painted Style edited by R. A. Schiffman, D. S. Whitley, A. P. Garfinkel, and S. B. Andrews, pp. 79-96. *Bakersfield College Publications in Archaeology*, Number 2.

Schiffman, R. A. and A. P. Garfinkel

1981a Prehistory of Kern County: An Overview. *Bakersfield College Publications in Archaeology*, Number 1.

1981b Obsidian Studies at the Ming Ranch Site (CA-Ker-983). In *Obsidian Dates III, A Compendium of the Obsidian Hydration Determinations Made at the UCLA Obsidian Hydration Laboratory*, C.W. Meighan and G. S. Russell, editors, pp. 125-129. Institute of Archaeology, University of California, Los Angeles, Monograph XVI.

Schiffman, R. A., D. S. Whitley, A. P. Garfinkel, and S. B. Andrews (editors)

1982 Pictographs of the Coso Region: Analysis and Interpretations of the Coso Painted Style. *Bakersfield College Publications in Archaeology*, Number 2.

Schroth, A. B.

1994 *The Pinto Point Controversy in the Western United States.* Ph.D. Dissertation, Department of Anthropology, University of California, Riverside.

Schultz, Peter D.

1979 Historical Faunal Remains from Panamint City: Notes on Diet and Status in a California Boom Town. *Pacific Coast Archaeological Society Quarterly* 15(4):55-63.

Scott, B. G. and T. M. O'Brien

1992 *Fish Timber Sale Archaeological Reconnaissance Survey, Cannell Meadow Ranger District.* Report on file, United States Department of Agriculture, United States Forest Service, Sequoia National Forest, Porterville, California.

Sennett-Graham, B.

1989 *Basketry: A Clue to Panamint Shoshone Culture in the Early 20th Century.* M.A. Thesis, Department of Anthropology, University of Nevada, Reno.

Shaul, D. L.

1986 Linguistic Adaptation and the Great Basin. *American Antiquity* 51(2):415-416.

Simms, S. R.

1985 Pine Nut Use in Three Great Basin Cases: Data, Theory, and a Fragmentary Material Record. *Journal of California and Great Basin Anthropology* 7(2):166-175.

Simpson, C. G.

1988 *Stylistic Boundaries Among Mobile Hunter-Foragers*. London: Smithsonian Institution Press.

Skinner, E. J.

1986 Appendix A: Analysis of Flaked Stone from the Crane Valley Hydroelectric Testing Program. In *Cultural Resources of the Crane Valley Hydroelectric Project Area* by S. K. Goldberg, S. S. Salzman, E. J. Skinner, J. Burton, M. E. Scully, J. J. Holson and M. J. Moratto, Volume III, Part 2, pp. 419-594. INFOTEC Research Inc. Report on file, Pacific Gas and Electric Company, San Francisco.

Slater, E.

2000 *Panamint Shoshone Basketry: An American Art Form*. Morongo Valley, California: Sagebrush Press.

Smith, C. R.

1978 Tubatulabal. In The Handbook of North American Indians, Volume 8, *California*, R. F. Heizer, editor, pp. 437-445. Washington, D.C.: Smithsonian Institution.

Smith, C. R. and D. Lee

2001 Pictographs of the Volcanic Tableland, Inyo and Mono Counties, California. In *American Indian Rock Art*, Volume 27, S. M. Freers and A. Woody, editors, pp. 201-210. American Rock Art Research Association.

Spears, J. R.

1892 *Illustrated Sketches of Death Valley*. Chicago and New York: Rand, McNally Co.

Stevens, N.

2001 *An Exploration of Temporal and Morphological Aspects of Concave Base Projectile Points from the Sierra Nevada and Western Great Basin*. Report on file, Department of Anthropology, California State University, Sacramento.

2003 Spatial and Temporal Patterning of Bedrock Mortar Sites in the Southern Sierra Nevada: A Regional Exploration. *Proceedings of the Society for California Archaeology* Volume 16: 175-182.

Steward, J. H.

1928 Pottery from Deep Springs Valley, Inyo County, California. *American Anthropologist* 30(2):348.

1929 Petroglyphs of California and Adjoining States. *University of California Contributions, Archaeological Research Facility* 24(2).

1933 Ethnography of the Owens Valley Paiute. *University of California Publications in American Archaeology and Ethnology* 33(3):233-350.

1937 Linguistic Distributions and Political Groups of the Great Basin Shoshoneans. *American Anthropologist* 39(4):625-634.

1938 Basin-Plateau Aboriginal Sociopolitical Groups. *Bureau of American Ethnology Bulletin* 120. Washington, D.C.

1941 Cultural Element Distributions, XIII: Nevada Shoshone. *University of California Anthropological Records* 4(2):209-360. Berkeley.

1968 Foreword. In *Rock Drawings of the Coso Range: An Ancient Sheep-hunting Cult Pictured in Desert Rock Carvings* by C. Grant, J. W. Baird and J. K Pringle, pp. vii-x. Maturango Museum Publication No. 4. Ridgecrest, California.

1970 The Foundations of Basin-Plateau Shoshonean Society. In *Languages and Cultures of Western North America*, E. H. Swanson, Jr., editor, pp. 113-151. Pocatello: Idaho State University Press.

Stine, S.

1990 Late Holocene Fluctuations of Mono Lake, Eastern California. *Palaeogeography, Palaeoclimatology, Palaeoecology* 78:333-381.

1994 Extreme and Persistent Drought in California and Patagonia during Mediaeval Times. *Nature* (369):546-549.

1995 Appendix F: Late Holocene Fluctuations of Owens Lake, Inyo County, California. In *Archaeological Evaluations of Thirteen Sites for the Ash Creek Project, Inyo County, California*, by A. J. Gilreath. Report on file, California Department of Transportation, Bishop, California.

Stoffle, R., L. Loendorf, D. Austin, D. Halmo, and A. Bullets

2000 Ghost Dancing the Grand Canyon: Southern Paiute Rock Art, Ceremony, and Cultural Landscapes. *Current Anthropology* 41:11-38.

Strong, E.

1960 Phoenix Buttons. *American Antiquity* 25(2):418-419.

1975 The Enigma of the Phoenix Button. *Journal of Historical Archaeology* 9:74-80.

Strong, M. F.

1971 *Desert Gem Trails. A Field Guide to the Gem and Mineral Localities of the Mojave and Colorado Deserts in California and Adjacent Areas of Nevada*. Mentone, California: Gembooks.

Stuiver, M. and P. J. Reimer

1993 *Radiocarbon Calibration Program* Rev. 4.4.2. Quaternary Isotope Laboratory, University of Washington, Seattle.

Stuiver, M., P. J. Reimer, E. Bard, J. W. Beck, G. S. Burr, K. A. Hughen, B. Kromer, F. G. McCormac, J. v.d. Plicht, and M. Spurk

1998 *Radiocarbon* 40:1041-1083.

Sutton, M. Q.

1981 Bighorn Sheep Rock Art from the Southern Sierra Nevada. *Masterkey* 55(1):13-17.

1986 Warfare and Expansion: An Ethnohistoric Perspective on the Numic Spread. *Journal of California and Great Basin Anthropology* 8(1):65-82.

1987 *A Consideration of the Numic Spread*. Ph.D. Dissertation, Department of Anthropology, University of California, Riverside.

1991 Approaches to Linguistic Prehistory. *North American Archaeologist* 12(4):303-324.

1994 The Numic Expansion as Seen from the Mojave Desert. In *Across the West: Human Population Movement and the Expansion of the Numa*, D. B. Madsen and D. Rhode, editors, pp. 133-140. Salt Lake City: University of Utah Press.

2001 Excavations at Teddy Bear Cave (CA-KER-508), Tomo-Kahni State Park, Southern Sierra Nevada, California. *Pacific Coast Archaeological Society Quarterly* 37(1):1-26.

Sutton, M. Q., S. R. Jackson, and F. A. Riddell

1994 Test Excavations at Seven Sites in the Southern Sierra Nevada near Lake Isabella, California. *Kern County Archaeological Society Journal* 5:22-85.

Swadesh, M.

1954 Time Depth of American Linguistic Groupings. *American Anthropologist* 56(3):361-364.

1959 Linguistics as an Instrument of Prehistory. *Southwestern Journal of Anthropology* 15:20-35.

Theodoratus Cultural Research, Inc. (TCR)

1984 *Cultural Resources Overview of the Southern Sierra Nevada*. Report on file, United States Forest Service, Sequoia National Forest, Kernville.

Thomas, D. H.

1971 *Subsistence-settlement Patterns in the Reese River Valley, Central Nevada*. Ph.D. Dissertation, Department of Anthropology, University of California, Davis.

1972 An Empirical Test of Steward's Model of Great Basin Settlement Patterns. *American Antiquity* 38(1):155-176.

1981 How to Classify the Projectile Points from Monitor Valley, Nevada. *Journal of California and Great Basin Anthropology* 3(1):7-43.

Thomas, D. H., L. S. A. Pendleton and S. C. Cappannari

1986 Western Shoshone. In Handbook of North American Indians, Volume 11, *Great Basin*, W. L. d'Azevedo, editor, pp. 262-283. Washington, D.C.: Smithsonian Institution.

Thomas, T.

1976 Petroglyph Distribution and the Hunting Hypothesis in the Central Great Basin. *Tebiwa* 18(2):65-74.

Thompson, R. S.

1900 Late Quaternary Vegetation and Climate in the Great Basin. In *Packrat Middens-The Last 40,000 Years of Biotic Change*, J. L. Betancourt, T. R. Van Devender and P. S. Martin, editors, pp. 200-239. Tucson: University of Arizona Press.

Titchenal, P. B.

1994 *A Chronology for Glass Beads from California and the Western Great Basin.* M.A. Thesis, Department of Anthropology, California State University, Chico.

Tuohy, D. R.

1984 Archaeology of Pyramid Lake, Western Nevada. *Nevada State Museum Anthropological Papers* 20.

Turner, J. L. and R. A. Weaver

1981 Water. In *The Desert Bighorn*, G. Monson and L. Sumner, editors, pp. 100-112. Tucson: The University of Arizona Press.

Turpin, S. A.

1990 Rock Art and Hunter-gatherer Archaeology: A Case Study from sw Texas and Northern Mexico. *Journal of Field Archaeology* 17:263-281.

Underwood, J.

2004 *Discovering the Desert Kawaiisu.* Manuscript in author's possession.

USDI (United States Department of the Interior)

2001 *United States Department of the Interior Fire Management Data CD-ROM. Weather Data File for Bear Peak and Walker Pass. Yearly Data Range 1976-1996.* Article electronically published and accessed at *http://famweb.usda.gov/kcfast/textfiles/wlstinv1044730.txt.*

Van Devender, T. R. and W. G. Spaulding

1979 Development of Vegetation and Climate in the Southwestern United States. *Science* 204(4394):701-710

Vaughan, S. J. and C. N. Warren

1987 Toward a Definition of Pinto Points. *Journal of California and Great Basin Anthropology* 9(2):199-213.

Voegelin, E. W.

1938 Tubatulabal Ethnography. *University of California Anthropological Records* 2(1):1-84. Berkeley.

von Werlhof, J. C.

1965 Rock Art of the Owens Valley, California. *University of California Archaeological Survey Reports* 65. Berkeley.

W & S Consultants

1999 *Class III Inventory/Limited Archaeological Testing Program for the Ducor Telephone Project, Kennedy Meadows, Tulare County, California.* Report on file United States Department of Agriculture, United States Forest Service, Sequoia National Forest, Porterville, California.

Walker, A. M.

1971 *The Rough and the Righteous of the Kern River Diggins.* Balboa Island, California: Poisson Press.

Wallace, W. J.

1965 A Cache of Unfired Clay Objects from Death Valley, California. *American Antiquity* 30(4):434-441.

Wallace, W. J., A. P. Hunt, and J. P. Redwine

1959 An Investigation of Some Stone Mounds in Death Valley National Monument. *Archaeological Research Associates Contributions to California Archaeology* 3.

Wallace, W. J. and E. T. Wallace

1978 *Ancient Peoples and Cultures of Death Valley National Monument.* Ramona, California: Acoma Books.

1979 *Desert Foragers and Hunters: Indians of the Death Valley Region.* Ramona, California: Acoma Books.

Walton, J.

1992 *Western Times and Water Wars: State, Culture and Rebellion in California.* University of California, Berkeley.

Warren, C. N.

1984 The Desert Region. In *California Archaeology* by M. J. Moratto, pp. 339-428. Orlando: Academic Press.

Warren C. N. and R. H. Crabtree

1986 Prehistory of the Southwestern Area. In Handbook of North American Indians, Volume 11, *Great Basin,* W. L. d'Azevedo, editor, pp. 183-193. Washington, D.C.: Smithsonian Institution.

Weissner, P.

1983 Style and Social Information in Kalahari San Projectile Points. *American Antiquity* 48(2):253-276.

1985 Style or Isochrestic Variation? A Reply to Sackett. *American Antiquity* 50(1):160-166.

Welles, R. E. and F. B. Welles

1961 *The Bighorn of Death Valley.* United States Department of the Interior, National Park Service, Washington, D.C.

Wells, H. F.

1983 *Historic and Prehistoric Pinyon Exploitation in the Grass Valley Region, Central Nevada: A Case Study in Cultural Continuity and Change.* Ph.D. Dissertation, Department of Anthropology, University of California, Riverside.

Whitley, D. S.

1982 *The Study of North American Rock Art: A Case Study from South-Central California.* Ph.D. Dissertation, Department of Anthropology, University of California, Los Angeles.

1988 Obsidian Hydration Dates from the Coso Range. In *Obsidian Dates IV: A Compendium of the Obsidian Hydration Determinations Made at the Obsidian Hydration Laboratory,* C. W. Meighan and J. L. Scalise, editors, pp 75-76 and 197-211. Monograph No. 29. Institute of Archaeology, University of California, Los Angeles.

1994 By the Hunter, for the Gatherer: Art, Social Relations and Subsistence Change in the Prehistoric Great Basin. *World Archaeology* 25(3):356-373.

1998 History and Prehistory of the Coso Range: The Native American Past on the Western Edge of the Great Basin. In *Coso Rock Art: A New Perspective,* E. Younkin, editor, pp. 29-68. Ridgecrest: Maturango Press.

Whitley, D. S. and R. I. Dorn

1984 Chemical and Micromorphological Analysis of Rock Art Pigments from the Western Great Basin. *Journal of World Archaeology* 6(3):48-51.

Whitley, D. S., R. A. Schiffman and S. B. Andrews

1982 Perspectives on the Painted Rock Art of the Coso Region. In Pictographs of the Coso Region: Analysis and Interpretations of the Coso Painted Style, R.A. Schiffman, D. S. Whitley, A. P. Garfinkel, and S. B. Andrews, editors, pp. 5-21. *Bakersfield College Publications in Archaeology,* Number 2.

Whitley, D. S., T. K. Whitley and J. M. Simon

2005 The Archaeology of Ayer's Rock (CA-INY-134), California. *Maturango Museum Publication* Number 19, Ridgecrest, California: Maturango Press.

Whitley, T.

1982a Coso Style Pictographs of CA-KER-735. In Pictographs of the Coso Region: Analysis and Interpretations of the Coso Painted Style, R. A. Schiffman, D. S. Whitley, A. P. Garfinkel, and S. B. Andrews, editors, pp. 22-46. *Bakersfield College Publications in Archaeology,* Number 2.

1982b Rock Art of CA-KER-736. In Pictographs of the Coso Region: Analysis and Interpretations of the Coso Painted Style, R.A. Schiffman, D. S. Whitley, A. P. Garfinkel, and S. B. Andrews, editors, pp. 47-66. *Bakersfield College Publications in Archaeology*, Number 2.

Wigand, P. E.

1997 Native American Diet and Environmental Contexts of the Holocene Revealed in the Pollen of Human Fecal Material. *Nevada Historical Society Quarterly* 40(1):105-116.

2002 Prehistoric Dynamics of Pinyon Woodland in the Northern Owens Valley Region and Beyond as Revealed in Macrofossils from Ancient Woodrat Middens. In *Draft Phase II Archaeological Investigations for the Sherwin Summit Rehabilitation Project, U.S. Highway 395, Inyo County, California* by J. Eerkens and J. King. Report on file, California Department of Transportation, District 6, Fresno.

Wigand, P. E., M. L. Hemphill, S. E. Sharpe, and S. Patra

1995 Great Basin Woodland Dynamics During the Holocene. In *Proceedings of the Workshop-Climate Change in the Four Corners and Adjacent Regions: Implications for Environmental Restoration and Land-Use Planning.* CONF-9409325, U.S. Department of Energy, Grand Junction, Colorado.

Williams, A.

2004 Subsistence Patterns in the Northwestern Mojave Desert: At Freeman Springs (CA-KER-6106). Paper presented at the Great Basin Anthropological Conference, Sparks, Nevada.

Wobst, H. M.

1974 Boundary Conditions for Paleolithic Social Systems: A Simulation Approach. *American Antiquity* 39(2):147-178.

Yohe, R. M., II

1992 *A Reevaluation of Western Great Basin Cultural Chronology and Evidence for the Timing of the Introduction of the Bow and Arrow to Eastern California Based on New Excavations at the Rose Spring Site (CA-Iny-372).* Ph.D. Dissertation, Anthropology Department, University of California, Riverside.

Yohe, R. M., II and M. Q. Sutton

1999 Settlement, Subsistence, Technology, Environmental Change, and the Rose Spring Period in the Western Mojave Desert. Paper presented at the 1999 Kelso Conference on California Desert Prehistory, Panamint Springs, California.

2000 Implications of Technological and Environmental Change in the Rose Spring Period in the Western Mojave Desert. Paper presented at the 2000 Annual Meetings of the Society for California Archaeology, Riverside, California.

Young, D. A. and R. L. Bettinger

1991 The Numic Spread: A Computer Simulation. *American Antiquity* 57(1):85-99.

Zeanah, D.

2002 Central Place Foraging and Prehistoric Pinyon Utilization in the Great Basin. In *Beyond Foraging and Collecting: Evolutionary Change in Hunter-Gatherer Settlement Systems*, pp. 231-256, B. Fitzhugh and J. H. Kluwer, editors. New York: Academic Publishers.

Zigmond, M.

1938 Kawaiisu Territory. In Tribal Distribution in the Great Basin, W. Z. Park (et al.), editors. *American Anthropologist* 40:634-638.

1941 *Ethnobotanical Studies Among California and Great Basin Shoshoneans.* Ph.D. Dissertation, Anthropology Department, Yale University.

1971 Kawaiisu Plant Name Categories. In Great Basin Anthropological Conference 1970: Selected Papers, C. M. Aikens, editor, pp. 155-165. *University of Oregon Anthropological Papers* 1. Eugene.

1972 Some Mythological and Supernatural Aspects of Kawaiisu Ethnography and Ethnobiology. In Great Basin Cultural Ecology: A Symposium, D. D. Fowler, editor, pp. 129-134. *Desert Research Institute Publications in the Social Sciences* No. 8. Reno.

1977 The Supernatural World of the Kawaiisu. In Flowers of the Wind: Papers on Ritual, Myth, and Symbolism in California and the Southwest, T. C. Blackburn, editor, pp. 59-95. *Ballena Press Anthropological Papers* 8. Socorro, New Mexico.

1978 Kawaiisu Basketry. *Journal of California Anthropology* 5(2):199-215.

1980 Kawaiisu Mythology, An Oral Tradition of South-central California. *Ballena Press Anthropological Papers* 18. Socorro, New Mexico.

1981 *Kawaiisu Ethnobotany*. Salt Lake City: University of Utah Press.

1986 Kawaiisu. In Handbook of North American Indians, Volume 11, *Great Basin*, W. L. d'Azevedo, editor, pp. 398-411. Washington, D. C.: Smithsonian Institution.

Zigmond, M. L., C. G. Booth, and P. Munro

1991 Kawaiisu, A Grammar and Dictionary with Texts, P. Munro, editor. *University of California Publications in Linguistics*, Volume 119. Berkeley: University of California Press.

Zimmerman, K. L., C. L. Pruett and M. Q. Sutton

1989 A Clovis-like Point from the Southern Sierra Nevada, California. *Journal of California and Great Basin Anthropology* 11(1):89-91.

Glossary

Anasazi: The ancestors of the modern Pueblo people of the American Southwest. A prehistoric culture centered around the Four Corners area and including parts of Arizona, Utah, Colorado, and New Mexico.

Assemblage: Refers to a collection or class of artifacts coming from a single site, for example the flaked-stone assemblage from the Rose Spring site.

Atlatl: A wooden throwing board used to extend the arm of a hunter in order to add strength and propulsion to the missile and thereby increase its accuracy and kill rate. Used in hunting principally large game such as artiodactyls. The implement is depicted in aboriginal rock drawings as a bisected circle. The transtition from the atlatl to the bow and arrow occurred in eastern California ca. AD 200/300 to 600.

Autochthonous: Indigenous, not exotic, development in place largely unaffected by significant outside influences.

Basal indentation: A characteristic element of either a dart or arrow point where there is a notch or larger concavity created by the manufacture, repair, or rejuvenation of the point at its bottom most part.

Biface: A flaked-stone artifact that has been chipped on both sides in an effort to reduce it to a more finished appearance and one that can readily be used to produce other stone tools or useful flakes or as a means to transport toolstone most efficiently.

Calendar years: See radiocarbon years.

Calibrated age: See radiocarbon years.

Catchment: The area traversed by hunter-gatherer groups during their seasonal rounds.

Coefficient of variation: Represents the ratio of the standard deviation to the mean, and it is a useful statistic for comparing the degree of variation from one data series to another, even if the means are drastically different from each other. A small CV of less than .25 for a sample of obsidian hydration measurements is indicative of a restricted, single period occupation. Larger measures (>.25) would be more representative of multiple occupations or lengthier periods of activity.

Comparative linguistics: The branch of historical linguistics concerned with comparing languages in order to identify and explain the historical relationships between them. The latter often involve reconstructing ancestral languages to explain how languages change and how they relate to one another.

Coprolites: The dessicated remains of human excrement, sometimes preserved in dry caves or other arid environments. Analyzing these residues has been found useful in developing a picture of the diet of prehistoric peoples. Recently it has also been discovered that coprolites preserve mitochondrial DNA and so may be used in the reconstruction of the genetic antecedents of aboriginal people.

Coso representational style petroglyphs: A style of rock drawing only found in Eastern California within the general vicinity of the Coso Range. These drawings are located largely within the confines of the China Lake Naval Air Weapons Station near Ridgecrest, California in eastern Kern and Inyo counties. The hallmarks of the style are a distinctive array of realistic renderings of bighorn sheep, medicine bags, patterned body animal humans, shields, and humans with weaponry (both bow and arrow and atlatl and dart) engaged in hunting campaigns. The most characteristic feature of the rock art style is the full front-facing, bifurcated horned, sheep, having a flat back and boat-shaped body with ears and hooves sometimes added. These petroglyphs were probably manufactured by a culturally distinct precursor population different from the late in-migration of Numic peoples. The petroglyphs appear to date mostly from ca. 2000 B.C. to AD 1300.

Coso style paintings: Rock pictures drawn with multicolored pigments found in eastern California in the former ethnographic territories of the Owens Valley Paiute, Panamint Shoshone (Koso), and Kawaiisu that depict men on horseback with European style hats, bovine animals (longhorn cattle?), and bighorn sheep. The pictures may have been part of a late/historic expression of Numic people associated with the ritual practices of the Ghost Dance. The sheep drawings are somewhat like Coso representational petroglyphs in that they include sheep with full front facing bifurcated horns, a hallmark of those earlier rock drawings. The paintings are probably copies of those more ancient petroglyph images by a later culturally distinct people.

Coso Volcanic Field: The region within eastern California situated within the confines of the China Lake Naval Air Weapons Station that includes the Coso Range north of Ridgecrest and south of Lone Pine. This area contains a number of sources of high quality volcanic glass used by the prehistoric occupants of the area known as Coso obsidian. This is also the region where the majority of the Coso Representational Style petroglyphs are located.

Debitage: Flaked stone debris created as a byproduct of reduction activities in the manufacture of stone tools, also known as waste flakes.

Dendogram: A tree diagram often used to show connections and relationships between various historically related entities. Such diagrams are often used to show the historical "family" relationships among languages.

Diagnostic point forms: Projectile points, darts, and thrusting spear types whose morphologies have been found to change through time and may be characteristic of a limited and defined time span.

Fluorescence: A period when an artifact or cultural expression peaks in popularity and appears to be represented most frequently in number and/or most elaborately in its expression.

Fremont: A distinct and unique prehistoric culture that inhabited the western Colorado Plateau and eastern Great Basin. They were a diverse group of hunters and farmers that are linked by a common style and technique of basketry, moccasin form, style of rock drawing and paintings, and a thin-walled grey pottery. The cultural expression of the Fremont is generally regarded as continuing from ca. AD 1 to 1500.

Game intercept: A hunting method involving the procurement of large game using natural topographic circumstances or diversion fences in a communal hunting strategy for the killing of bighorn, antelope or deer in the Great Basin.

Glottochronology: In linguistics, the technique used to estimate the time of divergence of two related languages. The method assumes that a basic vocabulary may be used as a sort of clock, on the assumption that languages change at a more-or-less constant rate through time. The method proposed a mathematical formula for establishing the date when two languages separated, based on a percentage of core or basic vocabulary of 100 (later 200) items that are the same (cognate) in the languages being compared. Morris Swadesh originally developed the technique and compiled a list of terms for this basic vocabulary. The basic vocabulary was composed of those terms common to every human language and not especially subject to cultural changes.

Granodiorite: An intrusive igneous rock similar to granite, but containing more plagioclase (pink) than potassium (white) feldspar. It usually contains abundant biotite mica and hornblende, giving it a darker appearance than true granite. Mica may be present in well-formed hexagonal crystals, and hornblende may appear as needlelike crystals.

Great Basin Scratched (also Numic Scratched): A petroglyph style presumed to have been manufactured by relatively recent occupants of the Desert West etching mainly abstract rock lines superimposed over or added to older (pre-Numic) rock drawings. The thin lines it is thought either were meant to embellish or deface the older glyphs.

Haplogroups: Branches on the dendogram or tree representing the genetic composition of a prehistoric population. These branches are defined by genetic mutations or "markers" found in Y chromosome and mitochondrial DNA. Haplogroups link members of a group back to the marker's first appearance, linking them with their most recent common ancestor. Haplogroups often have a geographic relationship.

Incurvate: A characteristic of some flaked-stone implements with special reference to dart or arrow points where the edge or margin is curved inward in a concave fashion.

Lacustrine: Characteristic of places that have excessive moisture near streams, lakes, rivers or marshes.

Lag quarry: A source of toolstone used by early prehistoric (before 3000 BP) peoples in the Coso Volcanic Fields away from the primary sources of large in-place seams of high quality volcanic glass.

Lanceolate: Shaped like a spear.

Lexical change: The extent to which a language differs from a related language belonging to the same family or having some evidence for an historical connection.

Linguistic isolate: In an absolute sense, a language that has no demonstrable genetic relationships with other languages. More commonly, used in a relative sense, to mean a language that does not belong to any established branch but instead forms its own independent branch of a language family. For example, Tubatulabal is a single member of its own branch when compared to other related language branches (as with the Numic languages) incorporated into the Uto-Aztecan linguistic stock.

Loanword analysis: An investigation into the character of the words from a particular language that appear to have been borrowed from neighboring languages. Such a study can be used to determine the direction and extent of cultural interactions.

Locus: A portion of an archaeological site that can be segregated and differentiated due to its distinct function and/or temporal placement.

Lovelock: A cultural expression typified by prehistoric materials discovered at Lovelock Cave and dating from ca. 3000 BC to AD 1000. The cultural assemblage in part includes duck decoys, wicker basketry, carved effigies, and fishing nets indicating a specialized emphasis on lakeside subsistence associated with the Humboldt marshlands.

Mano: A round or loaf-shaped stone used as a hand held implement to grind seeds on a stone milling slab to produce flour.

Medicine bag: A type of petroglyph image found in the Great Basin and American Southwest that resembles hide containers of talismans typically carried by aboriginal peoples as a means of increasing their success in life and in hunting. These images alternatively may represent full-body hunting disguises.

Medieval climatic anomaly (MCA): Several periods of epic drought that were exhibited roughly simultaneously in widely separated areas of the world during the period from ca. AD 900 to 1350.

For the southwestern Great Basin these periods of intense dryness have been reconstructed based on submerged tree stumps in Mono Lake, dendochronology on redwood trees (Sequoias) in the far southern Sierra Nevada, and other paleoclimatic data.

Mitochondrial DNA: DNA is present inside the nucleus in every cell of the human body but it is the DNA of the cell's mitochondria that is most commonly used to trace genetic ancestry. Mitochondrial DNA has been particularly useful with regard to reconstructing the history of human populations because it is inherited strictly through the mother and is present in large numbers in each cell. It may be preserved for thousands of years within the skeletal remains of ancient humans.

Monoconically drilled: A term used in describing the way a whole may be set in a shell or stone bead when manufactured only from a single side.

Numic hunting camps: Prehistoric aboriginal campsites found in eastern California dating to the period from ca. AD 600 to 1300 that indicate specialized emphasis on the mass harvest of easily procured small game animals including jack rabbits and grebes.

Pedoturbation: The mixing of soils caused by various agents. Faunal pedoturbations includes the effects of ants, earthworms, moles, and rodents. Human induced changes incorporate pedoturbation from tillage and other cultural activities that are ground disturbing actions.

Petroglyphs: Rock drawings often created on rock types where the lighter heart rock shows through the darker stained or desert varnished surface rock and in so doing creates an image of long standing. Petroglyphs are often manufactured (pecked) with a hard angular quartz/quartzite hammerstone into black-brown patinated basalt.

Pictographs: Painted rock pictures rendered with native pigments most commonly in hues of red, orange, yellow, black, and white.

Pre-contact: The time interval before Euro-American intrusions affected indigenous cultures in a particular region.

Pre-Numic: Prior to the late prehistoric intrusion of Numic peoples into eastern California and their migration across most of the Great Basin. The appellation Pre-Numic is intended to mean non-Numic and to identify a culturally, linguistically and perhaps genetically distinct precursor population antecedent to the Great Basin Shoshoneans or Numic peoples.

Primary outcrop: A location where high quality volcanic glass can be quarried associated with the largest, natural occurrence, of in-place obsidian toolstone. In this book the term refers to the obsidian quarry sites at Sugarloaf, West Sugarloaf, Cactus Peak, or Joshua Ridge within the Coso Volcanic Field.

Proto-vocabulary: A word list developed through comparison of different members of related languages. The reconstructed proto-vocabulary and languages is often proposed as a common ancestor for related languages.

Radiocarbon years: When radiocarbon dating was introduced, it was assumed that a radiocarbon year was equal to a calendar year. This was based on the assumption that the level of 14carbon in the atmosphere remained constant over time. It was later discovered that radiocarbon dates when compared with calendar year dates derived from ancient tree rings were not equivalent because atmospheric concentrations of 14carbon have fluctuated over time. The correction, or calibration, of radiocarbon dates younger than approximately 15,000 years can be accomplished by means of a calculation based on tree-ring dates. This calculation results in a "calibrated radiocarbon age" that is usually expressed as "cal BP" or the calibrated age.

Refugia: Places where Native Americans found refuge from hostile environments either from climatic conditions or conflict with Euroamericans.

Scoria: A heavy, dark, igneous rock that contains many bubblelike cavities. It is foamlike in appearance and occurs as a product of explosive eruptions and also as frothy crusts on some lavas. Scoria is sometimes called volcanic cinder and can be various colors including brown, black, and red.

Semiflexed and loosely flexed burials: Human interments that are bent in a generally fetal position or are somewhat more tightly constricted but not fully so. These orientations have cultural significance and are regulated by traditional religious practices or perhaps as a function of the ethnic identity of the departed person.

Shaman: A ritual adept, a healer or medicine man who often serves as a religious leader in forager communities.

Shamanic rituals: Religious observances led by a ritual adept and commonly found among hunter-gatherers.

Synchronicity: Dating to a similar or nearly identical time span.

Time-diagnostic artifacts: Manmade objects that have been found through archaeological study to change form over time and as such can be used as surrogate clocks to date associated prehistoric remains.

Typological affiliation: The association of an artifact or groups of artifacts with other similar forms having cultural, functional, technological, and/or temporal significance.

Vesicular basalt: A fine-grained, fire-formed (igneous) rock that is normally dark in color and contains many tiny holes (vesicles) that formed due to gas bubbles in the lava or magma. This type of rock is very porous, may resemble a sponge, and might even float on water.

Worter-und-Sachen Method: The words-and-things method originally developed in the study of Indo-European languages. The method uses a proto-vocabulary developed using the techniques of comparative linguistics, which attempt to reconstruct the history and cultural patterns of a region based on a reconstructed vocabulary.

Author Index

Subject Index